Nanostructured Materials and Nanotechnology VI

Nanostructured Materials and Nanotechnology VI

A Collection of Papers Presented at the 36th International Conference on Advanced Ceramics and Composites January 22–27, 2012 Daytona Beach, Florida

Edited by
Sanjay Mathur
Suprakas Sinha Ray

Volume Editors
Michael Halbig
Sanjay Mathur

The American Ceramic Society

WILEY

A John Wiley & Sons, Inc., Publication

Published by John Wiley & Sons, Inc., Hoboken, New Jersey.
Published simultaneously in Canada.

For general information on our other products and services or for technical support, please contact our
Customer Care Department within the United States at (800) 762-2974, outside the United States at
(317) 572-3993 or fax (317) 572-4002.

Wiley also publishes its books in a variety of electronic formats. Some content that appears in print may
not be available in electronic formats. For more information about Wiley products, visit our web site at
www.wiley.com.

Library of Congress Cataloging-in-Publication Data is available.

ISBN: 978-1-118-20597-6
ISSN: 0196-6219

Printed in the United States of America.

10 9 8 7 6 5 4 3 2 1

Contents

Preface

The 6th International Symposium on Nanostructured Materials and Nanotechnology was held during the 36th International Conference and Exposition on Advanced Ceramics and Composites, in Daytona Beach, Florida during January 22-27, 2012. This symposium provided, for the sixth consecutive year, an international forum for scientists, engineers, and technologists to discuss new developments in the field of nanotechnology. The symposium covered a broad perspective including synthesis, processing, modeling and structure-property correlations in nanomaterials and nanocomposites. Over 60 contributions (invited talks, oral presentations, and posters) were presented by participants from universities, research institutions, and industry, which offered interdisciplinary discussions indicating strong scientific and technological interest in the field of nanostructured systems. This issue contains 14 peer-reviewed papers cover various aspects and the latest developments related to processing of nanoscaled materials.

The editor wish to extend their gratitude and appreciation to all the authors for their cooperation and contributions, to all the participants and session chairs for their time and efforts, and to all the reviewers for their valuable comments and suggestions. Financial support from the Engineering Ceramic Division of The American Ceramic Society (ACerS) is gratefully acknowledged. The invaluable assistance of the ACerS staff of the meetings and publication departments, instrumental in the success of the symposium, is gratefully acknowledged,

We believe that this issue will serve as a useful reference for the researchers and technologists interested in science and technology of nanostructured materials and devices.

SANJAY MATHUR
University of Cologne
Cologne, Germany

SUPRAKAS SINHA RAY
National Centre for Nano Structured Materials
CSIR, Pretoria, South Africa

Introduction

This issue of the Ceramic Engineering and Science Proceedings (CESP) is one of nine issues that has been published based on content presented during the 36th International Conference on Advanced Ceramics and Composites (ICACC), held January 22–27, 2012 in Daytona Beach, Florida. ICACC is the most prominent international meeting in the area of advanced structural, functional, and nanoscopic ceramics, composites, and other emerging ceramic materials and technologies. This prestigious conference has been organized by The American Ceramic Society's (ACerS) Engineering Ceramics Division (ECD) since 1977.

The 36th ICACC hosted more than 1,000 attendees from 38 countries and had over 780 presentations. The topics ranged from ceramic nanomaterials to structural reliability of ceramic components which demonstrated the linkage between materials science developments at the atomic level and macro level structural applications. Papers addressed material, model, and component development and investigated the interrelations between the processing, properties, and microstructure of ceramic materials.

The conference was organized into the following symposia and focused sessions:

Symposium 1	Mechanical Behavior and Performance of Ceramics and Composites
Symposium 2	Advanced Ceramic Coatings for Structural, Environmental, and Functional Applications
Symposium 3	9th International Symposium on Solid Oxide Fuel Cells (SOFC): Materials, Science, and Technology
Symposium 4	Armor Ceramics
Symposium 5	Next Generation Bioceramics

Symposium 6	International Symposium on Ceramics for Electric Energy Generation, Storage, and Distribution
Symposium 7	6th International Symposium on Nanostructured Materials and Nanocomposites: Development and Applications
Symposium 8	6th International Symposium on Advanced Processing & Manufacturing Technologies (APMT) for Structural & Multifunctional Materials and Systems
Symposium 9	Porous Ceramics: Novel Developments and Applications
Symposium 10	Thermal Management Materials and Technologies
Symposium 11	Nanomaterials for Sensing Applications: From Fundamentals to Device Integration
Symposium 12	Materials for Extreme Environments: Ultrahigh Temperature Ceramics (UHTCs) and Nanolaminated Ternary Carbides and Nitrides (MAX Phases)
Symposium 13	Advanced Ceramics and Composites for Nuclear Applications
Symposium 14	Advanced Materials and Technologies for Rechargeable Batteries
Focused Session 1	Geopolymers, Inorganic Polymers, Hybrid Organic-Inorganic Polymer Materials
Focused Session 2	Computational Design, Modeling, Simulation and Characterization of Ceramics and Composites
Focused Session 3	Next Generation Technologies for Innovative Surface Coatings
Focused Session 4	Advanced (Ceramic) Materials and Processing for Photonics and Energy
Special Session	European Union – USA Engineering Ceramics Summit
Special Session	Global Young Investigators Forum

The proceedings papers from this conference will appear in nine issues of the 2012 Ceramic Engineering & Science Proceedings (CESP); Volume 33, Issues 2-10, 2012 as listed below.

- Mechanical Properties and Performance of Engineering Ceramics and Composites VII, CESP Volume 33, Issue 2 (includes papers from Symposium 1)
- Advanced Ceramic Coatings and Materials for Extreme Environments II, CESP Volume 33, Issue 3 (includes papers from Symposia 2 and 12 and Focused Session 3)
- Advances in Solid Oxide Fuel Cells VIII, CESP Volume 33, Issue 4 (includes papers from Symposium 3)
- Advances in Ceramic Armor VIII, CESP Volume 33, Issue 5 (includes papers from Symposium 4)

- Advances in Bioceramics and Porous Ceramics V, CESP Volume 33, Issue 6 (includes papers from Symposia 5 and 9)
- Nanostructured Materials and Nanotechnology VI, CESP Volume 33, Issue 7 (includes papers from Symposium 7)
- Advanced Processing and Manufacturing Technologies for Structural and Multifunctional Materials VI, CESP Volume 33, Issue 8 (includes papers from Symposium 8)
- Ceramic Materials for Energy Applications II, CESP Volume 33, Issue 9 (includes papers from Symposia 6, 13, and 14)
- Developments in Strategic Materials and Computational Design III, CESP Volume 33, Issue 10 (includes papers from Symposium 10 and from Focused Sessions 1, 2, and 4)

The organization of the Daytona Beach meeting and the publication of these proceedings were possible thanks to the professional staff of ACerS and the tireless dedication of many ECD members. We would especially like to express our sincere thanks to the symposia organizers, session chairs, presenters and conference attendees, for their efforts and enthusiastic participation in the vibrant and cutting-edge conference.

ACerS and the ECD invite you to attend the 37th International Conference on Advanced Ceramics and Composites (http://www.ceramics.org/daytona2013) January 27 to February 1, 2013 in Daytona Beach, Florida.

MICHAEL HALBIG AND SANJAY MATHUR
Volume Editors
July 2012

NANOSTRUCTURED COATINGS BY CLUSTER BEAM DEPOSITION: METHOD AND APPLICATIONS

Emanuele Barborini, Simone Vinati, and Roberta Carbone
Tethis SpA
Milan, Italy

ABSTRACT

One of the key issues to be addressed in order to exploit nanomaterial peculiar properties is the way devices, and surfaces in general, can be functionalized by nanomaterials. To allow the jump beyond lab-scale, deposition techniques are asked to fulfill requirements such as reliability and repeatability, batch deposition, scalability, compatibility with micromachining techniques.

Here we show how supersonic cluster beam deposition, based on pulsed microplasma cluster source, may answer these requests, while offering at the same time a wide library of available nanomaterials, including carbon, oxides, and noble metals. The growth of nanostructured functional coating takes place directly onto whatever surface exposed in front of the cluster beam. Cluster soft-assembling generates nanoporosity and, as a consequence, coatings with large specific surface, which are particularly suited for applications where interaction with liquid solutions or gas-phase atmospheres has to be favored.

Results on the integration of nanostructured coatings into devices with applicative purposes in sensing field and in biotech field, will be reported. They include gas sensing, stretch sensing, protein adsorption, cell adhesion, and selective capture of peptides, as examples of functions that may be added to devices by cluster beam deposition of nanomaterials.

INTRODUCTION

One of the most important ways to exploit the peculiar properties of nanomaterials is their integration into devices or generally onto surfaces, in form of functional coatings. For example, devices with applications in the field of energy storage/production, such as supercapacitors, fuel cells, or electrochemical photovoltaic cells, may enhance their performances by using high specific surface nanostructured coatings deposited onto their electrodes, in order to favor the interaction with the liquid or gas phases.[1-4] Usually, the synthesis of the precursors of the nanostructured coatings and the integration in devices are two well-separated production steps, whose characteristics may frequently limit the range of the applications of the nanomaterials from a given synthesis route. For example, the use of nanomaterials from wet-chemistry synthesis routes as active sensing layer into micromachined gas sensing platforms, is hampered on one side by the mechanical delicacy of micromachined parts, and on the other side by the presence of solvents in precursor material.[5-6] Moreover, nanomaterials from synthesis routes that require high temperature calcination step are incompatible with thermolable substrates, such as polymers. Finally, the functionalization of devices in batch requires patterned deposition with suitable lateral resolution, in order to deposit nanomaterials onto the proper functional area, within each single device. Although photolithography is the common approach to fulfill this task, it raises non-trivial issues in the case of micromachined devices.[6] Hence, a demand does exist of alternative methods for the production of nano-enhanced systems, where nanoparticle synthesis, manipulation, and integration steps are synergic parts of a unique process, overcoming most of the limitations of current approaches.

Here we describe a gas-phase method, based on supersonic cluster beams, for the functionalization of surfaces and devices with nanomaterials that may offer interesting opportunities with respect to nanomaterial integration issue. The first part of the paper will report on the deposition technique, with particular emphasis on cluster source. Results of morphological and structural characterization of cluster-assembled materials by means of electron microscopy and scanning probe techniques will be also reported. The second part of the paper will show some relevant examples of applications of nanostructured materials by supersonic cluster beam deposition within various applicative fields in sensing and biotech, such as chemical and mechanical sensing, protein adsorption, cell adhesion, and peptide selective capture.

EXPERIMENTAL

Supersonic Cluster Beam Deposition (SCBD)

As shown in figure 1, SCBD apparatus basically consists of two vacuum chambers, each having its own pumping system. The cluster source is connected to the first chamber, which is named expansion chamber. Driven by pressure difference between source inner and expansion chamber, a supersonic expansion of a inert gas (typically Argon), takes place through the source nozzle, carrying clusters out from production region (source inner), towards deposition region. At odd with respect to effusive sources, supersonic expansion causes the cluster beam to concentrate within a divergence of few degrees, ensuring that a large fraction of the material produced into the source is directed towards deposition region.

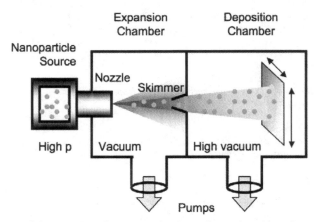

Figure 1. Scheme of the apparatus for supersonic cluster beam deposition. Arrows in deposition chamber indicate substrate rastering for large area depositions.

The second chamber is separated from the expansion chamber by means of a collimator with aerodynamic shape (skimmer), which removes most of the gas load due to cluster carrier, allowing only the central portion of the cluster beam to reach deposition region. In the case that the growth of cluster-assembled material should be performed in ultra-high-vacuum (UHV) conditions, as in Surface Science experiments, the number of vacuum chambers may be increased (according to the so-called differential vacuum approach) since beam collimation anyway provides for an adequate material collection.

If the coverage of areas exceeding the size of the cluster beam spot, which is typically few cm^2, is needed, substrate rastering may be adopted: exploiting motorized sample holder, substrate scanning in front of the cluster beam is operated, so that areas extending up to few hundreds of cm^2 can be processed.

Pulsed Microplasma Cluster Source (PMCS)

Although various cluster sources adopt supersonic expansion for the production of their cluster beam, such as Laser Vaporization Sources or Pulsed Arc Sources[7], we will focus here on Pulsed Microplasma Cluster Source (PMCS). PMCS is a rather recent system that we engineered and scaled-up in the last decade, and used to explore various applications of nanostructured coatings, as described in Sensing and Biotech Results section.

As first described in[8] and successively in[7-9], PMCS consists of a cylindrical ceramic element, hosting a suitable reaction cavity. A solenoid pulsed valve for the injection of the inert carrier gas closes the back side of the cavity, while a nozzle closes the front side. The solenoid valve is typically backed with a 50 bar gas line, and operates at an opening time of about two hundreds of microseconds. A channel perpendicular to cavity axis holds the metal rod-shaped target, which is used for clusters production. In presence of the inert gas pulse from solenoid valve, the metal target is negatively pulsed by means of a high-voltage, high-current dedicated power supply. An electric discharge from grounded nozzle to negatively-pulsed target takes place, generating a plasma of the inert gas. Plasma jet impinges on target surface and vaporizes part of it. Then, ablated atoms thermalize and condense to form clusters.

The pressure difference between the source cavity and the vacuum chamber where the source is faced (expansion chamber) causes the expansion across the source nozzle of the clusters-inert gas mixture, generating the supersonic clusters beam. The supersonic expansion accelerates clusters at a kinetic energy of few tenths of eV per atom, thus promoting an adequate adhesion of the resulting cluster-assembled coating while preserving the cluster original structure and avoiding any significant damage or heating of the substrate.

If supersonic expansion is forced through aerodynamic lenses[10], it is possible to obtain a highly collimated (divergence <1 deg) and intense cluster beam (aerodynamic focusing). By exploiting this property, patterned deposition of nanostructured coatings with sub-micrometric lateral resolution can be produced by non-contact stencil mask.[11] This feature marks a fundamental difference with respect to other gas-phase deposition techniques, since it allows for the easy integration of nanomaterials in functional areas of micro electro mechanical systems (MEMS) and micromachined platforms in general, avoiding photolithographic approach (see Gas sensing section below).

COATING CHARACTERIZATION

Atomic Force Microscopy

Surface morphology of cluster-assembled coatings has been studied by atomic force microscopy (AFM). Four examples are reported in figure 2, where the surface morphology of Ti, Hf, Zr, Fe nanostructured oxides is shown. Similar morphological features are observed in almost every nanostructured coating obtained by SCBD-PMCS. Therefore, we suppose that they can be ascribed to the deposition process in itself, and in particular to the low kinetic energy and limited diffusion of nanoparticles at the impact with the substrate, determining ballistic regime growth[7,12], which is characterized by nanoscale porosity, poorly-connected and non-compact structures with lower density respect to bulk and a surface roughness increasing with thickness.

AFM has been also adopted to evaluate the size distribution of coating precursors. To this purpose, very low coverage samples, with isolated nanoparticles, were deposited. Careful substrate preparation before deposition, as well as reference samples not exposed to the cluster beam, were adopted to favor artifacts identification/elimination in AFM images. Statistic of in-plane diameters was compared with statistic of heights, in order to identify the nanoparticle subset characterized by the height-diameter relation of spheroidal objects, and rule out non-spheroidal ones. In addition, size over-estimation due to AFM tip radius was avoided by limiting the counts of final size distribution to the heights of spheroidal objects subset. Size distributions were found to be lognormal, as expected for nanoparticles growth by gas-phase monomer aggregation into PMCS, as reported in[13] in the case of gold nanoparticles.

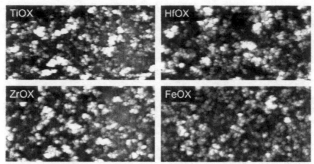

Figure 2. AFM images showing the surface morphology of four nanostructured oxides, with thicknesses around 200 nm. Each image has a size of 1×2 μm^2. It clearly appears that surface morphologies are very similar among the different oxides, a feature that is ascribable to growth dynamics of deposition process in itself.

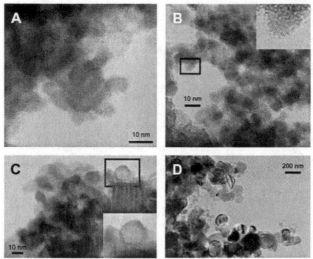

Figure 3. TEM images showing the nanostructure of as-deposited WO_3 film (A) and its evolution after thermal treatments at 200, 400, and 800 °C (B, C, D, respectively). Absence of lattice fringes inside nanoparticles of as-deposited film (A) indicates an amorphous structure. Annealing causes the evolution of nanoparticle structure toward crystalline order, which is almost completely reached after 400 °C. A progressive increase of crystal size is also observed, however nanoporous structure is preserved up to 800 °C. Reprinted with permission from[14].

Transmission Electron Microscopy

The as-deposited coatings have generally an amorphous and porous structure at the nanoscales, independently on the material. As in the case of surface morphology by AFM, this result may be attributed to the deposition process in itself.

In order to investigate the stability of the film nanostructure with respect to temperature, post-deposition annealing in pure air has been carried out. Annealing causes the amorphous grains to rearrange into a crystalline structure, while the nanoporosity is preserved up to temperature of several hundreds of °C. Figure 3 shows transmission electron microscopy (TEM) images of as-deposited, 200, 400, and 800 °C annealed tungsten oxide coating, chosen as a paradigmatic example.[14] Absence of lattice fringes inside nanoparticles of as-deposited film indicates for them an amorphous structure; size is about 10 nm. After 200 °C annealing, crystalline seeds nucleate within the amorphous aggregates and lattice fringes start becoming visible. After 400 °C, nanocrystals increase their size to about 20–30 nm, while a fraction of the material maintain its amorphous characteristic. After 800 °C, the average grain size increases to approximately 150 nm, while at higher temperatures (1000 °C, not shown) the cluster-assembled material loses its nanocrystalline feature and transforms to a polycrystalline continuous film.

It is worth to highlight that grain growth is minimal after annealing up to at least 400 °C, and that the nanoporous structure is preserved even at higher temperatures (800 °C). This is a fundamental feature for the use of cluster-assembled nanoporous materials as active layer in metal-oxide gas sensors, whose operation temperature is in the range 300–400 °C.

Spectro-ellipsometry

Beside microscopic techniques, SCBD-PMCS coatings have been studied with optical techniques, such as spectro-ellipsometry, as reported in[13]. Here, the optical response of cluster-assembled gold was modeled by explicitly introducing film porosity and finite-size effects due to nanoparticle size. Indications on clusters size as obtained by the optical model matched extremely well the size distribution obtained by AFM on isolated clusters. Since clusters are in contact each other into the coating, this result demonstrates that they retain their crystallographic individuality when assembled in three-dimensional structures.

The same kind of measurements were subsequently adopted to investigate the interaction between cluster-assembled nanoporous coatings and the liquid phase, at increasing degree of complexity of liquid compound/mixture: ethanol, octadecanethiol (C18), yeast cytochrome c (YCC).[15-17] A clear correlation between the density of the material and the fraction of pores that could be filled by ethanol molecules was observed. In the presence of abundant open pores, C18 molecules were observed to diffuse within the coating interior and bind to the pore walls, while in presence of porous coatings with less abundant open pores the tendency of the molecules to remain confined to the surface region, adopting a self-assembled monolayer (SAM) configuration.

YCC has been chosen as molecular system model to probe optical methods in detecting conformational changes upon adsorption at surfaces. Well-defined features related to molecular optical absorptions typical of the YCC heme group appear in the same position found for molecules in solution. This suggests that YCC native conformation is maintained upon adsorption onto nanostructured-nanoporous gold substrate.

SENSING AND BIOTECH RESULTS

Gas sensing

Since few decades, metal-oxide-based gas sensors are attracting considerable interest due to large potential impact on several applicative areas such as air quality monitoring, control of industrial processes, detection of harmful emissions, etc.[18-20] Metal-oxide gas sensors operate according to the change of the resistance of an oxide layer, occurring at high temperature in the presence of reactive compounds.[21-23] Due to their large specific surface area favouring the interaction with the atmosphere, nanoporous oxides can be successfully exploited as gas sensing active layers. With respect to other techniques, SCBD-PMCS method has the advantage of allowing direct integration of nanoporous metal-oxide layer onto platforms for sensing, at room temperature, with sub-micrometric lateral resolution, and avoiding any pre- or post-deposition treatment. Being a gas-phase deposition method, it provides the delicacy needed for functionalization of micromachined platforms, such as those based on the extremely delicate microhotplates.[24-26]

As reported in[27,28], we used SCBD-PMCS to integrate nanostrucutred oxide layer into microhotplate-based platforms for gas sensing (figure 4). Exploiting the high collimation of supersonic cluster beams, the deposition of nanomaterials was patterned with high lateral resolution by non-contact hard mask. This allows skipping photolithographic step, whose use may result as very difficult in the case of micromachined substrates. Gas sensing performances were characterized respect to various oxidizing and reducing species diluted in air, such as NO_2, ethanol vapors, and hydrogen. The measurements (not shown) suggest a detection limit in the 10-100 ppb range, linearity up to several tens of ppm, and fast response and recovery times. The microsensors were operated at temperatures in the range 200-300 °C spending as low as few tens mW of heating power.

As a perspective, the wide library of materials available by PMCS; the possibility to produce them with very similar morphological features, as shown in Coating Characterization section; and the capability to pattern the deposition, may together play a key role in advanced chemical sensing based on array approach. Beyond gas sensing, the overall features of SCBD-PMCS method may disclose new opportunities within the general issue of parallel integration of nanomaterials in MEMS.

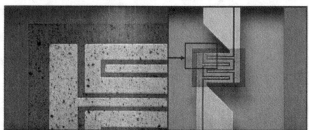

Figure 4. Optical microscope images showing a microhotplate-based platform for gas sensing after the deposition of nanostructured WO_3 coating (the darker rectangle in the centre of right image, enlarged in left image). The interdigitized electrodes pair and the serpentine-like heater, typical of metal-oxide gas sensors, are visible in right image. This platform is part of a batch of one hundred devices, which have been simultaneously deposited. Reprinted with permission from[27].

Mechanical stretch sensing

As reported in[29], nanocomposites with peculiar electrical properties can be produced by exposing a stretchable polymer, such as poly-dimethil-siloxane (PDMS), to supersonic cluster beam by PMCS. The kinetic energy of clusters, gained during supersonic expansion, is enough to induce a certain degree of penetration into the soft substrates, according to the so-called supersonic cluster beam implantation (SCBI) phenomenon, the authors report. A nanocomposite layer extending few tens of nanometers below the surface of the soft polymer is then generated, where clusters are embedded into polymer. Ianocomposite surface layer provides the polym er the capability of conducting an electrical current, whose intensity depends on the amount of percolating paths available for electron transport, among embedded nanoparticles.

As the polymer is stretched, the relative distance between embedded clusters increases, reducing the number of percolating paths, that in turns decreases electrical conduction. Once the polymer is released, original conduction is recovered. Proof of concept experiment exploiting a motorized uniaxial stretcher was reported in[29], where the reversibility of the phenomenon was demonstrated over 50,000 cycles of 40% applied strain. This result marks a dramatic difference with respect to the behavior of evaporated metallic films onto polymers, whose performances typically degenerate soon, mainly for the formation of cracks and for delamination.[30]

From a general point of view, results reported in[29] disclose the possibility to use supersonic cluster beam to easily create thin conductive patterns onto flexible substrates: a step ahead towards flexible electronics. Although the perspective of the research in[29] regards the production of flexible electrodes for biomedicine and smart prosthetics, nanocomposites by SCBI may offer the base transducing element for sensors of mechanical stretch.

Protein absorption

Within the technological trend reshaping biological analysis and assays towards high-content/high-throughput tools and methods, the development of novel miniaturized devices, such as biochips and microarrays, plays a key role. The first issue to be addressed in such devices, is the way how biological entities (DIA, proteins , cells) can be fixed in suitable positions, in stable and reliable configuration. This in turn depends on the optimization of the complex interactions that occur between biological entities and the immobilizing surface. In this framework, nanostructured oxides may be an interesting alternative to standard approaches to adhesion, in force of a nanostructure-mediated adhesion mechanism, which co-exists with other useful properties of the oxides, such as for example, the transparency at visible wavelength and the absence of auto-fluorescence.

As reported in[31], we have characterized nanostructured TiO_2 deposited on standard glass slides by SCBD-PMCS, as protein binding surface, in comparison with mostly diffused commercial substrates for protein and antibody microarrays. Ianostructured TiO_2 showed remarkable properties, in terms of protein adsorption, optical transparency at visible wavelengths (due to optical gap larger than 3 eV^{32}), absence of auto-fluorescence background, and signal-to-noise ratio, suggesting its possible use in different protein microarray applications. The compatibility of SCBD-PMCS deposition method with several substrates (glass, quartz, silicon, polymers), as well as with microfabrication techniques, as shown in the section on gas sensing, suggests that cluster-assembled nanostructured materials may represent a new family of functional coatings for immunodetection on miniaturized biochips and biosensors.

Cell adhesion

In the framework of the studies on the interaction between cluster-assembled nanostructured coatings and biological entities, we extended the research on cell adhesion in[33] to other materials and cell lines. We observed, as a general result, that nanostructured oxides by SCBD-PMCS, such as for example TiO_2, ZrO_2, WO_3, favor the adhesion of cell. This holds even in the case of living hematopoietic (circulating) cells, which are non-adhering cells by definition.

Ianostructured coatings with 50 nm thickness have been deposited on standard glass slides and annealed overnight at 250 °C, in clean and dry air atmosphere. After growing under standard conditions, living hematopoietic U937 model cells have been collected by centrifugation. Cellular pellets have been subsequently resuspended in Phosphate Buffer Saline (PBS) at 10^7 cells/ml.

To check cell adhesion in conditions of shear stress induced by moving flow, as those encountered in microfluidic systems, glass slides with straight, 300 μm wide and 50 μm depth, microfabricated channels have been clamped onto nanostructured oxide coated glass slides. The same setup has been used with standard Poly-D-Lysine coated slides (a common substrate for cell adhesion) and uncoated slides, as reference. 1.5 μl of cell suspension (total amount of about 15.000 cells) has been loaded into each channel and incubated for 2 minutes at 37 °C. After cell fixation with common fixative agents (paraformaldheide 4% or methanol/acetic acid 1:1) and extensive PBS washing, we have performed DAPI staining to visualize the nuclei. Then microfluidic slides have been disassembled and stained cells on coated and uncoated slides have been analyzed through fluorescence microscopy.

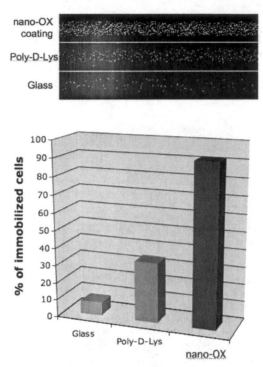

Figure 5. Results of cell adhesion experiments, in condition of microfluidic shear stress. In this case, nano-OX is TiO_2. Upper image shows cells (white dots) along microchannel at the end of one adhesion experiment. With respect to the total amount of cells injected into microchannel, at the end of the adhesion experiment we observed around 80% on nanostructured oxide coated slide, while less than 10% on bare glass slide (histogram graph).

Figure 5 shows an example of adhesion result after microfluidic experiment. Cells appear as a stable and homogeneous layer on nanostructured oxide coated slides, while on the control uncoated glass slides fewer cells are present. Poly-D-Lysine performance stays in between. We carried out several independent experiments to statistically evaluate the amount of cells surviving microfluidic shear stress test. We found that around 80% of cells are recovered at the end of the experimental protocol on nanostructured oxide coated slide, while less than 10% are present on the uncoated glass slide.

Optical transparency and good adhesive properties, even in the case of haematopoetic cells, make nanostructured oxides by SCBD-PMCS an interesting class of materials, which may be successfully adopted to solve the open issue of cell adhesion into microdevices for cytological applications with optical readout, as fluorescence-based analytical assays, and generally for lab-on-a-chip applications.[34]

Peptide selective capture

In the framework of biomarkers research, whose main applicative purpose is the identification of compounds helping in early diagnosis of diseases, Proteomics investigates the complex relationships between pathologies and the "Proteome", which is the huge ensemble of proteins within the organism. Among the biochemical processes occurring on Proteome, phosphorylation is a fundamental and reversible reaction modulated by kinase and phosphatase enzymes, whose action is frequently associated to pathological condition. In real clinical samples the identification and quantification of phosphorylated proteins is made particularly challenging, since it deals with the targeting of specific compounds dispersed at low concentration into high complexity samples. This task is usually addressed by Matrix-Assisted-Lased-Desorption-Ionization (MALDI) Mass Spectrometry, a powerful analytical technique providing high-sensitivity for detection of low concentration peptides.

Before undergoing MALDI analysis, a sample preparation step is required, where phosphorylated peptides are separated from complex samples, according to a standard enrichment method. This is a rather elaborated protocol, based on chromatographic approach, which exploits the affinity between phospho-peptides and metal-oxides, in form of small-size beads, pre-packed into a suitable chromatographic column.

Enrichment protocol can be largely simplified and shortened by performing phospho-peptide selective capture directly on MALDI plate, which has been previously functionalized with an oxide coating to this purpose. Although many techniques do exist to prepare an oxide-functionalized MALDI plate, the characteristics of SCBD-PMCS make it particularly suited for this purpose. In particular, the nanoporous nature of the coatings offers a large specific surface area for efficient interaction with liquid samples containing peptide mixtures, while the wide library of different oxides available allows to exploit the different level of specific affinity of various materials towards the variety of phospho-peptides.[35]

As reported in[36], we deposited Ti, Zr, Hf and Fe nanostructured oxides on MALDI plates and used them within experiment of phospho-peptide enrichment, in collaboration with MS Proteomics Unit at IFOM-IEO Campus. Figure 6 shows the comparison between two mass spectra, one collected from untreated MALDI plate and one from nanostructured TiO_2 coated plate, where selective capture of phospho-peptide is demonstrated through the signal of the singly phosphorylated 2061 peptide. Exploiting various concentrations of betacasein, a limit-of-detection of about 100 fmol was observed, which is quite similar to that of standard approach based on TiO_2 beads. Phopho-peptide selective capture was also observed in the case of complex samples, such as protein mixture (betacasein, fetuin, BSA, ovalbumin, ribonuclease A, cytochrome C) and real biological samples (Jdc80-Spc25 protein complex).

Exploiting the large library of nanostructured oxides by SCBD-PMCS, we envisage the possibility to adopt a combinatorial approach in the characterization of complex peptide samples in one single MALDI analysis session. This may be done by processing the same sample on an array-like MALDI plate, hosting various nanostructured oxides, and exploiting affinity differences between functionalized peptides and different oxides. In perspective, array-like MALDI plates by SCBD-PMCS could then become a novel tool for proteomic base research, as well as for clinical analysis of real samples.

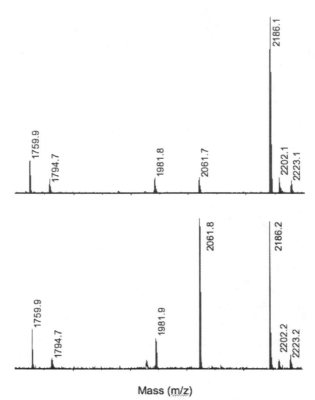

Mass (m/z)

Figure 6. MALDI-TOF mass spectra showing selective capture of phosphorilated peptides. The phenomenon has been observed through the comparison of the peak height of reference phosphorilated peptide (m/z = 2061) with peak height of reference non-phosphorilated peptide (m/z = 2186), whose intensity has been used for spectra normalization. Top spectrum has been obtained by processing a reference peptide mixture on standard MALDI plate without any functionalization; bottom spectrum has been obtained by processing the same peptide sample on nanostructured TiO_2 coated plate. Phosphorilated / non-phosphorilated peaks heights ratio of reference 2061-2186 peptides greatly increases if peptides sample is processed onto functionalized plate.

CONCLUSION

Supersonic Cluster Beam Deposition based on Pulsed Microplasma Cluster Source allows the direct integration of nanomaterials in devices, in form of a nanostructured and nanoporous coating. The characteristics of the method include room temperature processing, compatibility with any substrate (micromachined platforms and thermolable materials included), hard mask patterning at sub-micrometric lateral resolution, as well as a wide library of nanomaterials available. Cluster soft-assembling generates nanoporous materials with large specific surface area, particularly suited for application where the interaction with liquid-phase or gas-phase environment has to be favored, such as in gas sensing, as well as in liquid-mediated biotech applications.

Peculiar properties of nanostructured coatings by SCBD-PMCS have been investigated in various applicative fields: gas sensing, stretch sensing, protein adsorption, cell adhesion, and peptide selective capture. In particular, the method allows for parallel and safe functionalization of hundreds microhotplate-based platforms for gas sensing; while nanoparticle penetration and embedding into soft polymers allows for the production of nanocomposites with superior stability with respect to mechanical deformation. Regarding biotech area, although the study of the interaction between nanostructured coatings and biological entities (cells, proteins, viruses, DNA, etc.) is still in its infancy, solid results are already available, such as those regarding the improvement of cell adhesion by nanostructured oxides.

On the bases of the research outcomes reported in the present paper, the direct integration of nanomaterials in micromachined platforms for advanced physical and chemical sensing, as well as in microfluidic systems for the miniaturization of biological assay and for biosensing, may be the key technological issue addressed and solved by SCBD-PMCS, for innovative nano-enhanced systems.

ACKNOWLEDGEMENTS

We would like to thank all the researchers of the various areas, with whom we collaborated during the studies of the preset paper. In particular, we are grateful to A. Podestà and M. Indieri (CIMAINA and Physics Department, University of Milan) for AFM characterizations; C. Ducati (Material Science and Metallurgy, University of Cambridge) for TEM characterizations; F. Bisio (Physics Department, University of Genova) for spectro-ellipsometric measurements; and G. Grigorean (IFOM-IEO Campus, Milan) for peptide selective capture experiments.

We also thank all the engineers and technicians at Tethis SpA for invaluable support during experiments, in particular G. Bertolini, M. Franchi and M. Leccardi for nanomaterial depositions and gas sensing experiments; A. Zanardi, M. De Marni, S. Venturini and G. Calzaferri for biological experiments involving protein adsorption and cell adhesion.

We definitely thank P. Milani (CIMAINA and Physics Department, University of Milan) for having inspired and started most of the studies presented here.

REFERENCES

[1] L. Diederich, E. Barborini, P. Piseri, A. Podestà, P. Milani, A. Schneuwly, and R. Gallay, Supercapacitors based on nanostructured carbon electrodes grown by cluster beam deposition, *Applied Physics Letters*, **75**, 2662 (1999).

[2] B. Bongiorno, A. Podestà, L. Ravagnan, P. Piseri, P. Milani, C. Lenardi, S. Miglio, M. Bruzzi, and C. Ducati, Electronic properties and applications of cluster-assembled carbon films, *Journal of Materials Science: Materials in Electronics*, **17**, 427 (2006).

[3] E. K. Athanassiou, R. N. Grass, and W. J. Stark, Chemical Aerosol Engineering as a Novel Tool for Material Science: From Oxides to Salt and Metal Nanoparticles, *Aerosol Science and Technology*, **44**, 161 (2010).

[4] B. O'Regan and M. Grätzel, A low-cost, high-efficiency solar cell based on dye-sensitized colloidal TiO_2 films, *Nature*, **353**, 737 (1991).

[5] M. Graf, A. Gurlo, N. Barsan, U. Weimar, and A. Hierlemann, Microfabricated gas sensor systems with sensitive nanocrystalline metal-oxide films, *Journal of Nanoparticle Research*, **8**, 823 (2006).

[6] S. Mathur and M. A. Carpenter (Eds.), Metal Oxide Nanomaterials for Chemical Sensors, *Springer*, Berlin, to be published.

[7] P. Milani and S. Iannotta, Cluster Beam Synthesis of Nanostructured Materials, *Springer*, Berlin, 1999.

[8]   E. Barborini, P. Piseri, and P Milani, A pulsed microplasma source of high intensity supersonic carbon cluster beams, *Journal of Physics D: Applied Physics*, **32**, L105 (1999).

[9]   K. Wegner, P. Piseri, H. Vahedi Tafreshi, and P. Milani, Cluster beam deposition: a tool for nanoscale science and technology, *Journal of Physics D: Applied Physics*, **39**, R439 (2006).

[10] P. Liu, P.J. Ziemann, D.B. Kittelson, and P.H. McMurray, Generating particle beams of controlled dimensions and divergence, *Aerosol Science and Technology*, **22**, 293 (1995).

[11] E. Barborini, P. Piseri, A. Podestà, and P. Milani, Cluster beam microfabrication of patterns of three-dimensional nanostructured objects, *Applied Physics Letters*, **77**, 1059 (2000).

[12] A. L. Barabasi and H. E. Stanley, Fractal Concepts in Surface Growth, *Cambridge University Press*, Cambridge, 1995.

[13] F. Bisio, M. Palombo, M. Prato, O. Cavalleri, E. Barborini, S. Vinati, M. Franchi, L. Mattera, and M. Canepa, Optical properties of cluster-assembled nanoporous gold films, *Physical Review B*, **80**, 205428 (2009).

[14] E. Barborini, C. Ducati, M. Leccardi, G. Bertolini, P. Repetto, and P. Milani, Ianostructured Refractory Metal Oxide Films Produced by a Pulsed Microplasma Cluster Source as Active Layers in Microfabricated Gas Sensors, *Japanese Journal of Applied Physics*, **50**, 01AK01 (2011).

[15] F. Bisio, M. Prato, O. Cavalleri, E. Barborini, L. Mattera, and M. Canepa, Interaction of Liquids with Ianoporous Cluster Assembled Au Films, *Journal of Physical Chemistry*, **114**, 17591 (2010).

[16] F. Bisio, M. Prato, E. Barborini, and M. Canepa, Interaction of Alkanethiols with Ianoporous Cluster Assembled Au Films, *Langmuir*, **27**, 8371 (2011).

[17] C. Toccafondi, M. Prato, E. Barborini, S. Vinati, G. Maidecchi, A. Penco, O. Cavalleri, F. Bisio, and M. Canepa, Yeast Cytochrome c Monolayer on Flat and Ianostructured Gold Films Studied by UV–Vis Spectroscopic Ellipsometry, *BioIanoScience* , **1**, 210 (2011).

[18] I. Yamazoe, Towards innovati on of gas sensor technology, *Sensors Actuators B*, **108**, 2 (2005).

[19] J. Marek, H. P. Trah, Y. Suzuki, and I. Yokomori (Eds.), Sensors for Automotive Technology, *Wiley-VCH*, Weinheim, 2003, p. 562.

[20] D. Kohl, Function and applications of gas sensors, *Journal of Physics D: Applied Physics*, **34**, R125 (2001).

[21] G. Sberveglieri, Gas Sensors: Principles, Operation and Developments, *Springer*, Berlin, 1992.

[22] I. Barsan and U. Weimar, Conduction model of metal oxide gas sensors, *Journal of Electroceramics*, **7**, 143 (2001).

[23] A. Rothschild and Y. Komem, The effect of grain size on the sensitivity of nanocrystalline metal-oxide gas sensors, *Journal of Applied Physics*, **95**, 6374 (2004).

[24] J. Suehle, R. E. Cavicchi, M. Gaitan, and S. Semancik, Tin oxide gas sensor using micro-hotplate by CMOS technology and in-situ processing, *IEEE Electron Device Letters*, **14**, 118 (1993).

[25] R. E. Cavicchi, J. S. Suehle, P. Chaparala, K. G. Kreider, M. Gaitan, and S. Semancik, Microhotplate gas sensor, *Proc. Solid State Sensor and Actuator Workshop* (Hilton Head, SC), 53 (1994).

[26] K. D. Benkstein, C. J. Martinez, L. Guofeng, D. C. Meier, C. B. Montgomery, and S. Semancik, Integration of nanostructured materials with MEMS microhotplate platforms to enhance chemical sensor performance, *Journal of Ianoparticles Research* , **8**, 809 (2006).

[27] E. Barborini, S. Vinati, M. Leccardi, P. Repetto, G. Bertolini, O. Rorato, L. Lorenzelli, M. Decarli, V. Guarnieri, C. Ducati, and P. Milani, Batch fabrication of metal oxide sensors on micro-hotplates, *Journal of Micromechanics and Microengineering*, **18**, 055015 (2008).

[28] M. Decarli, L. Lorenzelli, V. Guarnieri, E. Barborini, S. Vinati, C. Ducati, and P. Milani, Integration of a technique for the deposition of nanostructured films with MEMS-based microfabrication technologies: application to micro gas sensors, *Microelectronic Engineering*, **86**, 1247 (2009).

[29] G. Corbelli, C. Ghisleri, M. Marelli, P. Milani, and L. Ravagnan, Highly Deformable Ianostructured Elastomeric Elect rodes With Improving Conductivity Upon Cyclical Stretching, *Advanced Materials*, **23**, 4504 (2011).

[30] I. M. Graz , D. P. J. Cotton, and S. P. Lacour, Extended cyclic uni-axial loading of stretchable gold thin-films on elastomeric substrate, *Applied Physics Letters*, **94**, 071902 (2009).

[31] R. Carbone, M. De Marni, A. Zanardi, S. Vinati, E. Barborini, L. Fornasari, and P. Milani, Characterization of cluster-assembled nanostructured titanium oxide coatings as substrates for protein arrays, *Analytical Biochemistry*, **394**, 7 (2009).

[32] E. Barborini, I. I. Kholmanov, A. M. Conti, P. Piseri, S. Vinati, P. Milani, and C. Ducati, Supersonic Cluster Beam Deposition of Ianostructured Titania, *European Physical Journal D*, **24**, 277 (2003).

[33] R. Carbone, I. Marangi, A. Zanardi, L. Giorgetti, E. Chierici, G. Berlanda, A. Podestà, F. Fiorentini, G. Bongiorno, P. Piseri, P. G. Pelicci, and P. Milani,•• Biocompatibility of cluster-assembled nanostructured TiO_2 with primary and cancer cells, *Biomaterials*, **27**, 3221 (2006).

[34] A. Zanardi, D. Bandiera, F. Bertolini, C.A. Corsini, G. Gregato, P. Milani, E. Barborini, and R. Carbone, Miniaturized FISH for Screening of Onco-hematological Malignancies, *Biotechniques*, **49**, 497 (2010).

[35] G. R. Blacken, M. Volný, M. Diener, K. E. Jackson, P. Ranjitkar, D. J. Maly, and F. Turecek, Reactive Landing of Gas-Phase Ions as a Tool for the Fabrication of Metal Oxide Surfaces for In Situ Phosphopeptide Enrichment, *Journal of American Society of Mass Spectrometry*, **20**, 915 (2009).

[36] P. Soffientini, A. Di Fonzo, M. Serra-Batiste, E. Barborini, R. Carbone, S. Vinati and G. Grigorean, FeOx, ZrOx, HfOx and TiOx MALDI plate nano-surfaces for selective capture of phosphorylated and sialylated peptides, *58th ASMS Conference on Mass Spectrometry*, Salt Lake City, 2010.

IN-SITU GROWTH OF CARBON NANOTUBES IN THREE DIMENSIONAL NEEDLE-PUNCHED CARBON FABRICS AND HYBRID ENHANCEMENT TO C/SiC COMPOSITES

Jianbao Hu [a,b,c], Shaoming Dong[a,b,*], Xiangyu Zhang[a,b], Bo Lu [a,b,c], Zhihui Hu [a,b,c], Jinshan Yang[a,b,c], Qinggang Li [a,b,c], Bin Wu, [a,b,c]
a State Key Laboratory of High Performance Ceramics and Superfine Microstructure, Shanghai Institute of Ceramics, Chinese Academy of Sciences, Shanghai 200050, China
b Structural Ceramics and Composites Engineering Research Center, Shanghai Institute of Ceramics, Chinese Academy of Sciences, Shanghai 200050, China
c Graduate University of Chinese Academy of Sciences, Beijing 100049, China

ABSTRACT

Multi-walled carbon nanotubes (CNTs) were grown by chemical vapor infiltration in three dimensional needle-punched carbon fiber felts. Systematical studies, by varying the reaction temperatures and the total pressure, were carried out in relation to morphology and quality of the products. The morphology of CNTs grown in the inner part of fiber felts were examined by scanning electron microscopy. Dense and uniform MWCNTs were drafted on the carbon fiber surface throughout the felt via controlling the reaction conditions. The composite reinforced with CNTs showed bundle pull-out mode due to enhancement of matrix intra bundles.

INTRODUCTION

Carbon nanotubes (CNTs) have attracted tremendous attention and interest due to their extraordinary unique properties. In addition to their outstanding mechanical properties[1], CNTs also possess excellent thermal and electrical properties[2], which make them attractive for reinforcing materials. By utilizing one or more properties of CNTs, multifunctional composites could be produced.

CNTs have been widely used as reinforcements in all kinds of composites, such as PMC, CMC and MMC. Although improvements have been discovered, the increase, especially for mechanical properties, are disappointing[3]. One reason is that the CNTs are easy to agglomerate and difficult to disperse uniformly in matrix. Compared with dispersing CNTs to matrix, CNTs directly grown on carbon fiber (CF) by chemical vapor deposition (CVD) is a more effective method to achieve uniform distribution and high load throughout a composite[4]. The CNTs growing on the fibers and extending into the matrix, are likely to enhance the matrix and provide additional fracture process, which may improve the toughness of material[5]. This method is a promising approach to achieve reinforcement by both nanoscale and microscale architecture.

The growth of CNTs on fiber by CVD is a mature technique, including the control of orientation, length and so on[6, 7]. The detail studies mainly focused on single fibers, two dimensional (2D) cloths and other fabrics[4, 7, 8]. There are seldom detail studies about growth of CNTs in three dimensional (3D) fiber fabrics. Actually, the growth of CNTs in 3D architecture is different from 2D architecture. The atmosphere of surface and inner parts are different. Therefore, it is necessary to study the growth conditions in order to achieve uniform growth of CNTs both on surface and inner of 3D fiber

---

* Corresponding author. Tel: +86 21 5241 4324; Fax: +86 21 5241 3903. Email address: smdong@mail.sic.ac.cn (Shaoming Dong)

architecture.

EXPERIMENTAL

Three-dimensional needle punched, 30vol.% carbon fiber felts, with dimensions $50\times40\times7mm^3$ were used here as substrate for carbon nanotube growth. These fiber fabrics were treated with HNO_3 for 5h to improve the adsorption of catalysts. The Fe/Al_2O_3 catalysts were introduced onto the modified fiber felts by immersing the felts into a ethanol solution of $Fe(NO_3)_3\cdot9H_2O/Al(NO_3)_3\cdot9H_2O$. The growth of CNTs was performed in a tubular quartz furnace (~45 mm in diameter) by using acetylene (C_2H_2) as hydrocarbon source in the chemical vapor deposition reaction. Hydrogen was also used as reaction gas. The pressure and reaction temperature on morphology of CNTs grown in carbon fiber felts were investigated. The deposition time were carried out for 60Min.

These carbon fiber felts grafted with CNTs were then densified using polymer impregnation pyrolysis (PIP) with polycarbosilane as precursors to fabricate C_f/SiC composites

The inner parts which located in the middle of the felts after the growth of CNTs were cut down using a knife. The inner layers of carbon fibers, were inspected using scanning electron microscopy (SEM) to characterize the morphology and distribution of CNTs. The fracture faces of C/SiC composites were inspected by SEM.

RESULTS AND CONCLUSION

Influence of Reaction Temperature

The deposition temperature is an important parameter which strongly affects the yield and morphology of CNTs. The influences of deposition temperature on the morphology of CNTs grown on the inner plane were shown in Fig. 2. When the reaction temperature is 650°C, the inner plane of carbon fiber felt can be covered with CNTs. Dense CNTs were also observed when the temperature reached 750°C. The length of CNTs was longer than those grown at 650°C. When reaction temperature was 850°C, carbon particles rather than CNTs were deposited on fibers. Fig. 2d shows the Raman spectra of CNTs grown at 650°C and 750°C. The Raman spectrum of products was not performed because no CNTs were grown on fibers at 850°C. The intensity ratio I_D/I_G for D band and G band is widely used for characterizing the defect quantity in graphitic materials[9]. The intensity ratios I_D/I_G of CNTs synthesized at 650°C and 750°C were 0.86 and 0.46. Higher ratio corresponds to higher defects.

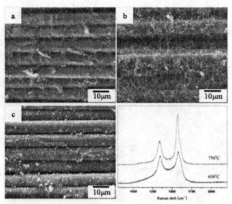

Figure 1. Morphology of CNTs grown at 650°C (a), 750°C (b) and 850°C (c). Raman spectra are shown in d.

The growth of CNTs in the three-dimensional architecture is quite different from those on surface of one-dimensional or two-dimensional architecture using chemical vapor deposition. It refers to the chemical vapor infiltration process. In addition to the homogeneous gas phase pyrolysis of carbon source and heterogeneous reaction (growth of CNTs) processes, the transportation of gaseous species inside a fibrous perform[10] should also be considered. For the heterogeneous reaction case, the activity of catalyst particles is relatively low at lower deposition temperature (such as 650°C), which causes lower growth rate and faster catalyst poisoning[7, 11]. As a consequence, low catalyst activity and faster poisoning resulted in shorter length of CNTs and more defects (Fig. 1a and Fig. 1d). At 750°C, the catalyst activity is increased and the catalyst poisoning effect is delayed thus more effective growth is achieved which leads to a longer growth of CNTs and less defects (Fig. 1b and Fig. 1d). On the contrary, at a higher temperature the catalyst particles tend to agglomerate together and formed bigger particles, which decreases the overall activity. On the other hand, excess pyrolysis of C_2H_2 would produce larger hydrocarbon species such as polycyclic aromatic hydrocarbons (PAHs)[10]. More PAHs would form during the infiltration into the inner part of fiber felt. Excess PAHs can lead to premature poisoning of catalysts and formation of soot. Therefore, pyrolytic carbon was produced in the inner plane of felt at 850°C.

Influence of The Total Pressure

The influence of deposition pressure on CNTs morphology on inner fiber plane is shown in Fig. 2. It clearly appears that an increase in the total pressure brings remarkable improvement in yield, morphology and coverage. When the pressure is about 1.5KPa, no CNTs were found. Dense CNTs were observed when the total pressure increased up to 3KPa. Better growth of CNTs can be found when the total pressure is 10KPa. The coverage is much more uniform and the length of CNTs is larger than those grown at 3KPa. Fig. 2d shows the Raman spectra of CNTs grown at 3KPa and 10KPa. The Raman spectrum of products was not performed because no CNTs were grown on fibers at 1.5KPa. The intensity ratios I_D/I_G of CNTs obtained at 3KPa and 10KPa were 0.46 and 0.50. The defects of CNTs grown at 3KPa are less than those grown at 10KPa, but the difference is unconspicuous.

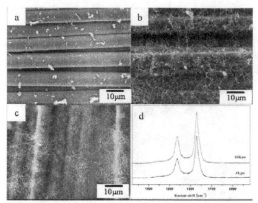

Figure 2. Morphology of CNTs grown at 1.5KPa (a), 3KPa (b) and 10KPa (c). Raman spectra are shown in d.

The substrates are three-dimensional porous structures. The infiltration of carbon source is needed. When the total pressure is low, C_2H_2 was entirely consumed during the growth of CNTs located at the surface of felts. No more C_2H_2 can infiltrate into the inner part. Therefore, no CNTs were grown at 1.5KPa. When the total pressure increased, the concentration of C_2H_2 rose up. Sufficient carbon source can reach the inner part of fiber felt and CNTs can grow successfully.

Hybrid Enhancement to C_f/SiC Composites

Fig. 3 shows the fracture surfaces of C_f/SiC Composites with and without CNTs. We can see that, there are different fracture modes between the composites with and without CNTs. Bundle pull-out were observed in composites with CNTs, while single fiber pull-out were shown in composites without CNTs. CNTs grown intra-bundles are likely to enhance the matrix among fibers in bundle, which caused the fracture of whole bundle. On the contrary, the composite without CNTs showed a normal fiber pull-out phenomenon.

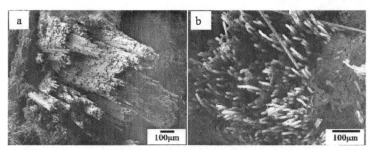

Figure 3 Fracture surfaces of C_f/SiC composites with (a) and without (b) CNTs.

CONCLUSION

CNTs can be grown in the three-dimensional needle punched carbon fiber felt via changing the total pressure and reaction temperature in chemical vapor deposition process. The increase of pressure improved the coverage of CNTs. The reaction temperature should carefully be controlled. Higher temperature promotes the formation of carbon particles rather than CNTs. The composite reinforced with CNTs showed bundle pull-out mode due to enhancement of matrix intra bundles.

ACKNOWLEDGMENTS

Author appreciate the financial support of the national natural science foundation of china under the grant No. of 51002170 and Innovation Program of Shanghai Institute of Ceramics Chinese Academy of Sciences under grant No. of Y12ZC6160G.

REFERENCES

[1]R.S Ruoff and D.C. Lorents, Mechanical and thermal properties of carbon nanotubes. *Carbon*, 33,925-930 (1995).

[2]T.W. Bebbesen, H.J. Lezec, H. Hiura, J.W. Bennett, H.F. Ghaemi, T. Thio, Electrical conductivity of individual carbon nanotubes. *Nature*, 382,54-56 (1996).

[3]S.S. Samal and B. Smrutisikha, Carbon nanotubes reinforce Ceramic Matrix composites-A Review, *J. Miner. Mater. Character. Eng.*, 7, 355-370 (2008).

[4]E.T. Thostenson, W.Z. Li, D.Z. Wang, R.F. Ren, T.W. Chou, Carbon nanotube/Carbon fiber hybrid multiscale composites, *J. Appl. Phys.*, 91, 6034-6037 (2002).

[5]H. Qian, A. Bismarck, E.S. Greenhalgh, M.S.P. Shaffer. Carbon nanotube grafted carbon fibres: A study of wetting and fibre fragmentation, *Composites: Part A* ,41, 1107-1114 (2010).

[6]N. Yamamoto, A.J. Hart, E.J. Garcia, S.S. Wicks, H.M. Duong, A.H. Slocum, B.L. Wardle, High-yield growth and morphology control of aligned carbon nanotubes on ceramic fibers for multifunctional enhancement of structural composites, *Carbon*, 47, 551-560 (2009).

[7]S.P. Sharma, S.C. Lakkad, Morphology study of carbon nanospecies grown on carbon fibers by thermal CVD technique, *Surf. Coat. Techno.*, 203, 1329-1335 (2009).

[8]R.B. Mathur, S. Chatterjee, B.P. Singh, Growth of carbon nanotubes on carbon fiber substrates to produce hybrid/phenolic composites with improved mechanical properties, *Compos. Sci. Technol.*, 68, 1608-1615 (2008).

[9]M.S. Dresselhaus, G. Dresselhaus. R. Saito, A. Jorio, Raman spectroscopy of carbon nanotubes. *Phys. Reports* , 409, 47-99 (2005).

[10]K. Norinaga, O. Deutschmann, Detailed Kenetic Modeling of Gas-Phase Reactions in the Chemical Vapor Deposition of Carbon from Light Hydrocarbons, *Ind. Eng. Chem. Res.*, 46, 3547-3557 (2007).

[11]Z.G. Zhang, L.J. Ci, J.B. Bai, The growth of multi-walled carbon nanotubes with different morphologies on carbon fibers, *Carbon*, 43, 633-655 (2005).

BORON NITRIDE NANOTUBES GROWN ON COMMERCIAL SILICON CARBIDE FIBER TOW AND FABRIC

Janet Hurst
Ching-cheh Hung
NASA Glenn Research Center
Cleveland Ohio

ABSTRACT

Continuous fiber-reinforced ceramic composites (CFCCs) have been considered for a variety of high temperature structural applications including within hot sections of gas turbine engines. SiC_f/SiC_m compositions are of particular interest for these turbomachinery applications. Boron Nitride Nanotubes (BNNT) are of interest as reinforcement for high temperature materials. In this work, the potential for reinforcement of SiC_f/SiC_m with BNNT is examined. Successful infiltration of both tows and performs was demonstrated. The impact of growing BNNT on SiC fiber tensile properties was also investigated. It was found that BNNT reinforcement resulted in greatly increased tensile strength for hot pressed $SiC_f/BNNT$ composites.

INTRODUCTION

The promise of nanotechnology is being translated into property improvements among many material systems. While significant advances have occurred for carbon nanotube (CNT) reinforced materials such as polymer matrix composites (PMC), the thermo-oxidative stability of carbon is insufficient for many aerospace applications. As an alternative to carbon, the use of Boron Nitride Nanotubes (BNNT) provides promise for surviving in these extreme environments. BNNT has mechanical properties which are similar to carbon nanotubes [1-3] but with superior temperature capability. [4-5] Silicon carbide fiber reinforced silicon carbide matrix composites (SiC_f/SiC_m) have been thoroughly investigated as a structural material for hot sections in gas turbine engines. [6-9] Incorporation of thermally resistant BNNT into SiC_f/SiC_m composites may enable improved mechanical properties. Previous work has demonstrated successful BNNT reinforcement within a glass composite [10] which resulted in improved tensile strength and fracture toughness. And for low temperature applications, BNNT has been successfully used to reinforce a thin film polystrene composite. [11] However material systems of great interest for propulsion applications such as SiC_f/SiC_m have not been examined.

In this work, the concept of "fuzzy" fibers, previously demonstrated by growing carbon nanotubes on the surface of carbon fibers, [12-14] was examined for BNNT grown on SiC fibers. Other researchers have found that this concept improved interlaminar shear strength properties by 69% for a CNT-thermoset resin system. [15] A schematic of this concept is shown in Figure 1. Similar results for ceramic composites may encourage utilization of laminate or 2D weaves rather resorting to difficult 3D approaches, perhaps justifying the expense of incorporating BNNT. Presented here is an exploratory effort in which BNNT was grown in a radial orientation on the surface of Hi-Nicalon™, Hi-Nicalon-S™, and Sylramic™ fibers. The objective of the work was to demonstrate BNNT growth on SiC surfaces and determine whether full infiltrate of tows and preforms was possible. Also another goal was to determine the effect of BNNT deposition on room temperature mechanical properties of SiC tows and preforms.

Figure 1. Schematic of "Fuzzy Fiber"; deposition of BNNT on SiC fibers.

EXPERIMENTAL PROCEDURE

Sections of SiC fiber tow, 9 cm in length, were processed in batches by an atmospheric chemical vapor deposition process previously developed at NASA GRC. This process uses nitrogen/ammonia gas mixtures and boron vapor from boron/B_2O_3 powder mixtures. Temperatures can be from 1100 to 1400°C. Initially only 1 or 2 tow lengths were infiltrated at a time but eventually up to 15 tow lengths were coated with no negative impact on BNNT deposition. There was little variation in BNNT throughout the lengths of the tows; the variation that occurred was concentrated in the last ½ -1 cm ends of the tows. While not strictly required for growth, catalysts are often utilized to maximize BNNT growth.[16,17] For this initial effort, a Fe containing catalyst was used at concentrations to maximize BNNT production levels while also minimizing potential damage to the fibers. SiC tows were composed of approximately 800 fibers for Sylramic tows and 500 filaments/tow for other products. Each SiC fibers within tows were approximately 8-10 μm in diameter. Additional woven preforms of Hi-Nicalon and Sylramic fibers were also infiltrated by using the same process. These specimens were cut from woven cloth into 2 cm by 10 cm specimens. Finally, rigidized preform samples of the same dimensions were infiltrated. Rigidized preforms were specimens which had a 1-2 μm layer of melt-infiltrated SiC deposited on the woven fiber. This permitted the specimens to maintain shape within the deposition reactor without sagging and also avoided tows becoming unraveled at the edges of the preform cloth. Following infiltration with BNNT, rigidized preform specimens were further cut into smaller pieces and assembled into 4-ply composites of 1.5 cm by 7 cm and hot pressed at 2000 C and 6 ksi in argon. Table 1 summarizes the samples prepared for this study. All samples were examined with a Hitachi S4700 Field Emission Scanning Electron Microscope (FeSEM) with an Energy Dispersive Spectrometer (EDS). Fracture load was measured for uncoated and BNNT-coated tows and tensile strength was measured for BNNT/SiC composites. An electromechanical Instron 5500 series load frame was used according to ASTM standard C1275 for composites and D3822 for tows. Gage lengths were 25 mm.

RESULTS

BNNT was successfully grown on each SiC surface studied in this work. Both SiC fiber tows and woven performs were successfully infiltrated with substantial quantities of in-situ grown BNNT. A summary of the types of samples synthesized in this effort is shown in Table 1. BNNT deposition and infiltration at 3 levels of architecture were studied. First infiltration and deposition within 3 types of SiC tows was successfully demonstrated. Next, SiC 8-harness satin (HS) fabric was infiltrated with BNNT. Finally, rigidized 8HS pinweave SiC preforms were infiltrated with BNNT and assembled into 4-ply composite coupons by hot pressing. Tensile behavior was examined at both the tow level and the composite level. Substantial improvement in tensile strength was demonstrated at the composite level. The initial focus of this work had been to demonstrate BNNT grown on SiC fiber tow outer surfaces with the goal of eventually providing improved interlaminar composite properties. However, it was shown that BNNT could fully infiltrate tows and fabrics, potentially providing composite strengthening.

Specimen type	Purpose	Status
Tows. In-situ growth and infiltration of BNNT on 3 types of SiC.	Feasibility of BNNT growth and infiltration on SiC tows.	Demonstrated abundant BNNT growth and infiltration of tows.
Fabric. In-situ growth and infiltration of SiC fabric. 3 runs evaluated infiltration parameters.	Feasibility of infiltration of woven material	Infiltration demonstrated throughout fabric. Heavy growth on outside of fabric, lighter on interior fiber surfaces.
Composite coupons. BNNT grown on and within rigidized SiC performs then hot pressed into 4 ply 2D SiC/BNNT composite coupons.	Initial composite mechanical property evaluation. Rigidized preform protects fiber surface from BNNT growth if necessary and eliminates sagging of preform.	2 1/2 - 3 x improvement in tensile strength with addition of BNNT.

Table 1. Summary of specimens.

Tow Level

Three different commercial SiC fiber tows were used as substrates for deposition of BNNT. Substantial BNNT growth was observed on all surfaces as well as within each tow. Some examples of this growth are shown in Figures 2-4. Growth was found to be very reproducible with several deposition runs made for each fiber type. Tows were completely encompassed with blankets of abundant nanotube growth, obscuring individual fiber filaments within the tows. Sections of tow were weighed before and following each deposition run. It was found that tows of Sylramic and Hi-Nicalon fiber types doubled in weight, shown in Table 2. Growth on these tows was similar with long nanotubes up to many millimeters in length which were uniformly deposited along the entire length. The appearances of Hi-Nicalon-S tows were different than the other two types. Each Hi-Nicalon-S tow was also completely covered with BNNT growth but with a reduced mass. At the onset of this project, a target nanotube diameter for in-situ growth had been selected as 100-150 nm based on the authors experience with dispersing BNNT in ceramic and glass matrices.[10] This was generally achieved with both Sylramic and Hi-Nicalon tows, particularly near the surfaces of tows where nanotube diameters primarily ranged from 80-150 μm in diameter, with a few individuals from 20 up to 200 μm. Hi-Nicalon-S tows exhibited finer BNNT diameters, 40-80 nm. BNNT was also deposited on fiber surfaces within interior tows and these nanotubes typically had smaller diameters, more typically 20-50 nm. This suggests that BNNT diameter and length is controlled by availability of vapor species. Constrained growth conditions resulted in shorter nanotubes with smaller diameters as shown in Figure 4. It was also found that spreading tows prior to BNNT deposition allowed increased nanotube length, diameter and nucleation density. The chemistry among the SiC fiber types varied. Sylramic fiber was a nearly stoichiometric SiC material with a C-rich surface layer. Small amounts of Ti and B have been shown to migrate to the surface, as is done in the fabrication of IBN-Sylramic fiber.[8] This may have provided additional nucleation sites for BNNT growth. Hi-Nicalon fiber has a Si-C-O surface and also developed a very large amount of BNNT growth. In contrast, Hi-Nicalon-S fiber tows, shown in Figure 4, while possessing the least amount of coating also had a Si-C surface that was slightly rich in carbon. The results of BNNT deposition of SiC tows are summarized in Table 2.

	BNNT Coatings on SiC tows									
Tow Substrate	Tow surface chemistry	NT rowth	NT length (on tow surface)	NT diameter (on tow surface)	NT length (within tows)	NT diameter (within tows – with no tow spreading)	BNNT(g)/SiC tow (g)(15-20 tows averaged)	As-received Fracture Load of Tows (N)(3-6 tows averaged	BNNT Coated Fracture Load Of Tows(N)(3-6 tows averaged)	Retained Strength %
Sylramic	Stoichiometric fiber. C-rich surface	Abundant	100's μms - several mm	80-200 nm	10-100 μm	20-50 nm	1.0	130	95	~75%
Hi-Nicalon-S	Stoichiometric fiber. Slightly C-rich surface	Fewer than Syl or Hi-Nic	100's μm	40-80 nm	5-50 μm	20-50nm	0.3	100	93	~100%
Hi-Nicalon	Si-C-O fiber and surface	Abundant	100's μms to several mm	80-200 nm	10-100 μm	20-50nm	1.1	120	65	~50%

Table 2. Summary of nanotube deposition results on SiC tows.

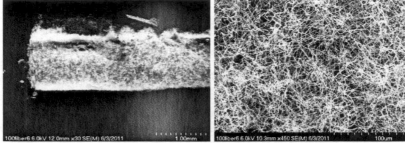

Figure 2. Sylramic SiC tows were completely covered with thick blankets of BNNT growth.

Figure 3. Hi Nicalon SiC tows were completely covered with thick blankets of BNNT growth.

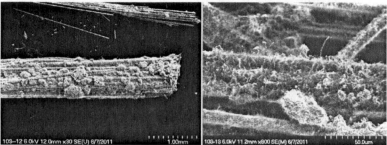

Figure 4. Coated Hi-Nic-S tow demonstrated bnnt growth but at a reduced level relative to Sylramic and Hi-Nicalon.

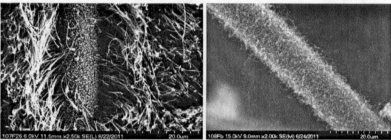

Figure 5 Individual fibers are coated within tows.

The adherence of BNNT was examined by immersing tow sections of each fiber type in an ultrasonic bath for 10 minutes in 20 ml of ethanol. Nanotubes were not removed from the SiC fiber surfaces although the ethanol became slightly cloudy, indicating that some nanotubes were removed. About 1/3 of the BNNT by weight was lost during this treatment. Figure 6 shows photos of BNNT coated Sylramic tow following this ultrasonic bath treatment. Tow surfaces remained densely coated with BNNT.

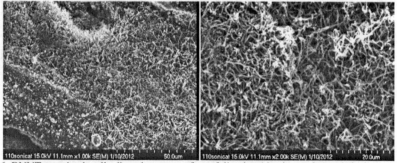

Figure 6. BNNT remained well adhered to tow surfaces following 10 minute ultrasonic bath.

Fracture loads for uncoated and BNNT-coated tows were also investigated as degradation of fiber strength by catalyst and/or BNNT growth was considered a possibility. Single filament strength testing of individual fibers was not possible as entangled nanotubes made it impossible to remove single filaments from tows. Two interesting points were noted during testing. First, the BNNT nanotubes were very well adhered to the fibers within the tows as demonstrated by both the previously discussed ultrasonic bath treatment and by observation of the fracture surfaces. Fracture surfaces were viewed following tensile testing. There was no spallation or observable loss of BNNT coating, as shown in Figure 7. Secondly, while fracture loads were diminished relative to uncoated specimens for Hi-Nicalon by about ½, and Sylramic tows by about 1/4, observations of the fractured surfaces of the BNNT-coated fibers did not reveal any pitting or damage to the fibers. The heavy BNNT coating reduced the handeablilty of tows significantly due to both the high modulus and volume of BNNT. In the case of typical planar BN coatings, only thin BN coating of 0.5 µm or so are produced and therefore bending strength or handeability of the fiber tows were not appreciably affected. The dimension and weight of the tows were often doubled with BNNT coatings grown on the fibers. This lead to some tows breaking when the load frame grips were closed prior to testing. Additionally, fiber tows were often not straight but had a slight permanent curvature resulting from two sources; being wound on a spool and sagging of the preform within the deposition chamber. This bend was then coated with heavy nanotube coatings, leading to substantial bending moment in the samples. In subsequent efforts, a "window frame" fixture will be used during BNNT growth to fabricate straight tows. The Hi-Nicalon-S tows, with much less BNNT coating, retained most of its load carrying capability.

Figure 7. BNNT remained well adhered to tow surfaces during fracture

Figure 8. BNNT coated exterior of preform and interior fiber surface within preform fabric

Preform and Composite Coupon Level –

BNNT was also grown throughout several Hi-Nicalon 8HS preforms. The results confirmed that BNNT infiltration of woven samples or even components is feasible in a reactor of appropriate dimensions. An example of a coated perform is shown in Figure 8. A heavy "blanket" of BNNT was deposited on the perform exterior with additional lighter growth within the preform. However these preform samples proved difficult to handle as unraveling of the preform edges occurred as well as sagging throughout the specimen. As this would cause uncertainty in the tensile strength results, another specimen approach was selected to investigate mechanical properties. Sylramic SiC 2-D pinweave rigidized preforms were chosen to infiltrate with in-situ grown BNNT and then fabricated into SiC_f/BNNT composites. Fabric both with and without BNNT was hot pressed at 2000°C and 6 ksi for 1 hour in argon to fabricate 4-ply 2-D composites with dimensions of 1.25 cm wide by 0.055 cm by 10 cm length. The only matrix phase present was a 1-2 μm rigidizing melt-infiltrated SiC layer on the perform surfaces and the in-situ grown BNNT. As the effect of BNNT growth on fiber surfaces could not be clearly determined from previous tow testing, this rigidizing layer may protect the fiber surface from potential damage during BNNT growth. Without a matrix phase, densities of the composites were low, 1.18 g/ cm^3 for composites with BNNT and 0.9 g/cm^3 for those without BNNT. Nevertheless, hot pressed composite samples were robust and easily handled. Following hot pressing, no discernable damage to the BNNT was observed, Figure 9. Substantial BNNT growth was observed on at least half of the weave surfaces with some growth on most surfaces. Composite coupons were tensile tested, with results shown in Figure 10. While these porous composites were weak, a 2 ½ - 3x improvement in tensile strength was observed for samples reinforced with BNNT. Both coupons without BNNT reinforcement failed at 4 MPa while those with BNNT failed at 10 and 13 MPa. This tensile strength result is likely due to nanotubes causing a reduction in crack initiation. A failed BNNT composite specimen is shown in Figure 11.

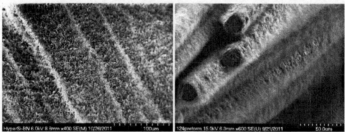

Figure 9. BNNT coating was retained on all fiber surfaces following hot pressing. More than half of individual fiber surfaces from the interior ply sections were fully covered. Additionally there was a very thick covering on the outer portions of plies.

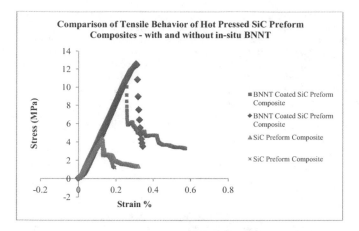

Figure 10. Addition of BNNT resulted in greatly improved tensile strength for SiC composite samples.

Figure 11. BNNT/SiC composite sample following tensile testing. Specimens demonstrated graceful composite failure.

CONCLUSION

In-situ growth of boron nitride nanotubes has been successfully accomplished on lengths of three commonly utilized advanced SiC fiber tows. Full infiltration of the tows and preforms with BNNT was demonstrated with ample BNNT growth. The load carrying capability of both Sylramic and Hi-Nicalon tows was somewhat diminished but improved test specimens will be fabricated to continue this investigation. Hi-Nicalon-S tows, with less BNNT growth, retained their full load carrying ability.

A significant result of this effort was the successful infiltration of SiC preform fabric. Also SiC 2-D woven performs were infiltrated to fabricate simple SiC/BNNT composites. Porous SiC composite coupons fabricated with BNNT reinforcement achieved 3 times the tensile strength relative to coupons without BNNT reinforcement. The BNNT survived very high temperature hot pressing conditions. This suggests that BNNT reinforcement in SiC composites may offer significant mechanical property improvements for high stress applications.

In the next phase of this work, dense SiC/BNNT composites will be fabricated to undergo thermomechanical testing including interlaminar shear strength evaluation.

ACKNOWLEDGEMENT
The authors would like to thank the NASA Glenn Center Innovation Fund for their support of this research.

REFERENCES
1. N.G. Chopra, A. Zettl, Solid State Commun. 105 (1998) 297.
2. X. Wei, M.S. Wang, Y. Bando, D. Golberg, D. Adv.Mater. 22 (2010) 489.
3. N. Chopra and A. Zettl, Solid State Commun., 105 [5] (1998) 297.
4. J. Hurst et al, Developments in Advanced Ceramics and Composites 26 [8] (2005) 355.
5. Y. Chen, et al. Appl. Phys. Lett, 84(13) (2004) p. 2430-2432.
6. K. Luthra, High Temperature Ceramic Matrix Composites for Gas Turbine Applications, (2001) 744 ISBN3-527-605622.
7. T. Oku, I. Narita, H. Tokoro,, J. Phys. Chem. Solids 2006, 67 (5-6), 1152.
8. H. Yun, J. DiCarlo, R.Bhatt, and J. Hurst, Ceram. Eng. Sci., 24 [3] (2003) p. 247.
9. R. Bhatt, J. Gyekenyesi, and J. Hurst, Ceramic Eng. Sci. 21 [3] (2000) p. 331.
10. N. Bansal, J. Hurst, S. Choi, J. Am. Ceram. Soc., 89 [1] (2006).
11. C. Zhi, Y. Bando, C. Tang, S. Honda, H. Kuwahara and D. Golberg, J. Mat. Res., 10 (2006).
12. Namiko Yamamoto et al, NanoEngineered Aerospace Structures Consortium, 2011.
13. K. Liew, Nanotechnology 22 (2011) 085701 (7pp)
14. E. J. Garcia, B.L. Wardle, A. J. Hart and N. Yamamoto, Carbon 28, (2010)
15. E. J. Garcia, B. L. Wardle, A. J. Hart and N. Yamamoto, Comp Sci Tech 68, [9], (2008).
16. L.T. Chadderton, Y. Chen, J. Cryst. Growth 2002, 240, 164-169.
17. Z.G. Chen, J. Zou, F. Li, G. Liu, D. Tang, D. Li, C. Liu, X. Ma, H. Cheng, G. Lu. Adv. Funct. Mater. 17 (2007) p.3371.
18. Oku, T.; Narita, I.; Tokoro, H. J. Phys. Chem. Solids 67 [5-6] (2006) 1152.
19. G. Chollen, R. Pailler, R. Naslain, F. Laanani, M. Monthioux and P. Olry, J. Mater. Sc.i 32 [2] (1997) 327.

A NEW GREENER SYNTHETIC ROUTE TO CADMIUM/LEAD SELENIDE AND TELLURIDE NANOPARTICLES

Neerish Revaprasadu
Department of Chemistry, University of Zululand, Private bag X1001, KwaDlangezwa, 3880.

ABSTRACT

CdSe, CdTe, PbSe and PbTe nanoparticles have been synthesized using a simple, non-organometallic hybrid solution based, high temperature route. The route involves reacting selenium or tellurium powder with sodium borohydride ($NaBH_4$) to produce selenide or telluride ions; followed by the reaction with a metal salt in a coordinating solvent. The metal source had considerable influence on the final morphology of the particles. The particles were characterised by X-ray diffraction and electron microscopy techniques.

INTRODUCTION

Cadmium and lead based chalcogenide nanoparticles have been at the forefront of nanomaterials synthesis and applications during the past decade. Chemically grown CdSe nanoparticles are probably the most extensively investigated colloidal II-VI semiconductor nanoparticles because of its unique size dependent chemical and physical properties which render it useful in electronic and biological applications.[1] Much of the work on CdSe has led to scientists having a better understanding of the electronic properties of semiconductor nanoparticles. The high photoluminescence (PL) quantum efficiencies of CdTe makes it an interesting material for use in applications such as light emitting devices,[2] photovoltaic and photoelectrochemical devices[3] and biological labels.[4] Recently there has also been considerable interest in lead based nanomaterials. The stable and tunable emission of NIR–emitting lead chalcogenide quantum dots (QDs) such as PbSe and PbTe make them suitable for applications in telecommunications (1300-1600 nm), bio-imaging (near IR-tissue window 800 and 1100 nm) and solar cells (800-2000 nm).[5]

The early synthetic routes to CdSe were sophisticated requiring specialised apparatus as it involved toxic, noxious and pyrophoric required starting materials at elevated temperatures. In particular the 'hot injection' method pioneered by Bawendi required the use of dimethylcadmium which is extremely hazardous.[6] The single molecular precursor route involved synthesizing diselenocarbamato complexes, which while overcoming the disadvantage of using dimethylcadmium involved the use of equally toxic carbon diselenide.[7] The search for alternative greener routes led to the modification of the Bawendi route as reported by Peng and Peng, where CdO replaced dimethyl cadmium as the precursor.[8] The authors have done extensive work on the shape control of CdSe nanoparticles by varying key reaction parameters such as temperature, precursor concentration and capping groups.[8,9]

CdTe has also been synthesized by the 'hot injection' method. Highly luminescent CdTe nanoparticles have been synthesized by reacting dimethylcadmium with tellurium sources in mixtures of dodecylamine and tri-octylphosphine.[10] Peng used his greener route to synthesise CdTe using CdO as a precursor instead of $Cd(CH_3)_2$.[8] The use of water soluble thiol based ligands to stabilise CdTe nanoparticle is an alternative to the organometallic route.[11,12] This route is generally simple, cheap and yield particles that have higher quantum yields than the organically passivated particles.

Lead chalcogenide nanocrystals, exhibit strong quantum size effect due to the large Bohr radii of both electron and holes which lead to enhanced quantum confinement effects. Various morphologies of PbSe nanocrystals have been reported including spheres,[13] cubes,[14] rings,[15] and wires.[16] The shape evolution of PbSe can be influenced by temperature, growth time, solvent and precursor delivery. The synthesis of PbSe nanocrystals has often involved the injection of a solution of a lead salt and tri-octylphosphine selenide (TOPSe) into a hot solvent,[15] thermolysis of single source precursors,[17] or hydrothermal synthesis.[18] Anisotropic PbTe nanoparticles is more challenging with only a handful reports of nanorods or tubes.[19]

This paper reviews a recently reported novel route to metal selenide and telluride nanoparticles. Good quality, organically passivated CdSe, CdTe, PbSe and PbTe have been synthesized using a simple, non-organometallic route that is environmentally friendlier than most reported routes to selenium and tellurium based nanoparticles.[20-24] The method involves reacting selenium or tellurium powder with sodium borohydride (NaBH₄) to produce the selenide or telluride ions; followed by reaction with a cadmium or lead salt. Variation of the metal salts and reaction temperatures produced particles with varying morphologies. The methodology could be used as a simple, cheaper route to selenium and tellurium based nanoparticles.

EXPERIMENTAL

CHEMICALS

Cadmium chloride, cadmium carbonate, lead chloride, lead carbonate, sodium borohydride (NaBH₄), deionised water (HPLC grade), methanol, toluene, hexadecylamine (HDA), tri-n-octylphosphine (TOP).tri-n-octylphosphine oxide, (TOPO technical grade 90 %) were purchased from Aldrich. Selenium and tellurium powder was purchased from Merck. All the chemicals were used as purchased. Solvents were distilled prior to use.

SYNTHESIS OF CdSe NANOPARTICLES

In a typical room temperature reaction, selenium powder was reduced to selenide ion by adding 0.32 mmol of selenium powder to 20.0 mL of HPLC grade water in a three neck flask. Sodium borohydride (0.80 mmol) was added to the flask contents and the flask was purged with nitrogen flow to create an inert atmosphere. The reduction reaction was carried out for 2 h with continuous stirring. A 0.32 mmol solution of cadmium salt (CdCl₂ or CdCO₃) was added to the reduced selenide ion solution. A light yellow to brown coloured solution was obtained depending on the cadmium source used. Stirring was continued for 5 min, followed by the addition of excess methanol to flocculate the CdSe particles. The resulting precipitate was isolated by centrifugation and dispersed in 15.0 mL of TOP. The resultant CdSe-TOP mixture was then injected into 6.0 g of TOPO (pre-heated to 230 °C) or HDA (pre-heated to 160 °C). After injection a drop (20 - 30 °C) in the temperature was observed. The temperature was then stabilized at the injection temperature and the reaction was allowed to proceed for one hour. After cooling, excess methanol was added to flocculate the nanoparticles. The flocculate was separated from the supernatant by centrifugation and the excess methanol was removed under vacuum to give TOPO or HDA-capped CdSe nanoparticles. The resultant particles were dissolved in toluene to give an optically reddish brown solution of nanoparticles for characterization.

SYNTHESIS OF CdTe NANOPARTICLES

Tellurium powder (0.32 mmol) was mixed with deionised water (20.0 mL) in a three-necked flask. Sodium borohydride (0.80 mmol) was carefully added to this mixture and the flask was immediately purged with nitrogen gas to create an inert atmosphere. After 2 h, 0.32 mmol of the CdCl₂ was dissolved in deionised water (20 mL) and added to the light grey tellurium ion solution. The solution was stirred for 30 min followed by the addition of excess methanol. The resultant solution was then centrifuged. The CdTe produced was dispersed in tri-n-octylphosphine, TOP (6.0 mL) and stirred continuously to form a TOP-CdTe solution, which was then injected into hot hexadecylamine, HDA (6.0 g) at 190°C. A sudden decrease in temperature was observed. An immediate addition of methanol resulted in the reversible flocculation of the nanoparticles. The flocculate was separated from the supernatant by centrifugation. The resultant particles were dissolved in toluene to give a solution of nanocrystallites for characterization.

The above reaction procedure was repeated using different cadmium sources ie. $Cd(CH_3COO)_2$ $Cd(NO_3)_2$; $CdCO_3$ at reaction injection temperature of 230 °C.

SYNTHESIS OF PbSe NANOPARTICLES

Selenium powder (0.32 mmol) was mixed with 20.0 mL of deionised water in a three neck flask. A 0.80 mmol of $NaBH_4$ was carefully added to this reaction mixture and the flask was immediately purged with nitrogen to facilitate an inert atmosphere. After 2 h, 0.31 mmol of lead salt, $PbCO_3$ added to the reaction mixture. The suspension was stirred for 30 min followed by the addition of excess methanol. The resultant suspension was then centrifuged. The precipitate was dispersed in TOP and injected into hot hexadecylamine at 190 °C and the reaction was allowed to continue for 4 h. The black solution was cooled to 65 °C. Addition of excess anhydrous methanol to the solution resulted HDA capped PbSe nanoparticles.

The above procedure was repeated at a reaction injection temperature of 230 °C. Another reaction was carried out using $Pb(NO_3)_2$ at 230 °C.

SYNTHESIS OF PbTe NANOPARTICLES

Tellurium powder (0.31 mmol) was mixed with 20 mL of deionised water in a three neck flask. 0.80 mmol of $NaBH_4$ was carefully added to this reaction mixture and the flask was immediately purged with nitrogen to facilitate an inert atmosphere. After 1.5 h, 0.31 mmol of $PbCO_3$ added to the reaction mixture. The suspension was stirred for 30 min followed by the addition of excess methanol. The resultant suspension was then centrifuged. The precipitate was dispersed in TOP and injected into hot hexadecylamine at 230 °C and the reaction was allowed to continue for 4h. The black solution was cooled to 65 °C. Addition of excess anhydrous methanol to the solution resulted HDA capped PbTe nanoparticles.

The above reaction procedure was repeated at a injection temperature of 270 °C. Two further reactions were carried out using $Pb(NO_3)_2$ and $PbCl_2$ as the lead sources at 230 °C.

INSTRUMENTATION
OPTICAL CHARACTERIZATION

A Cary 50 conc. UV-Vis Spectrometer was used to carry out optical measurements in the 200-800 nm wavelength range at room temperature. Samples were placed in quartz cuvettes (1 cm path length).

Room-temperature photoluminescence (PL) spectra were recorded on a Perkin Elmer LS 55 luminescence spectrometer with Xenon lamp over 200-800 nm range. The samples were placed in quartz cuvettes (1 cm path length). All optical measurements were carried out at room temperature under ambient conditions.

X-RAY POWDER DIFFRACTION

The crystallinity of the dried colloids was studied by using powder X-ray diffractometer (XRD). Powder diffraction patterns were recorded in the high angle 2θ range of 5-80° using a Bruker AXS D8 Advance X-Ray diffractometer, equipped with nickel filtered Co $K\alpha$ radiation ($\lambda = 1.5418$ Å) at 40 kV, 40 mA and at room temperature. The scan speed sizes were 0.05° min^{-1} and 0.00657 respectively.

ELECTRON MICROSCOPY (TEM AND HRTEM)
Samples were prepared by placing a drop of dilute solution of nanoparticles on Formvar-coated grids (150 mesh). The samples were allowed to dry completely at room temperature and viewed using a JEOL 1010 TEM and JEOL 2100 HRTEM. Viewing was done at an accelerating voltage of 100 kV (TEM) and 200 kV (HRTEM), and images captured digitally using a Megaview III camera; stored and measured using Soft Imaging Systems iTEM software.

RESULTS AND DISCUSSION
SYNTHETIC METHODOLOGY
The method employed in this study is a modification of Rao *et.al.*[25] who prepared t-selenium and tellurium nanorods and nanowires in water using selenide or telluride ions as the source of selenium or tellurium. In our reaction, colloidal cadmium and lead selenide and telluride nanoparticles have been synthesized by the addition of an aqueous cadmium salt to a freshly prepared oxygen-free NaHSe/Te. The isolated solid product was dispersed into TOP and injected into hot HDA at temperatures of 190, 230 or 270 °C and held at the same temperature for 2-4 h. After cooling to 50 °C the nanoparticles were isolated by the addition of methanol to the reaction mixture. The sequence of reactions is shown in below equations.

$$4NaBH_4 + 2Se/Te + 7H_2O \longrightarrow 2NaHSe/Te + Na_2B_4O_7 + 14H_2 \quad\quad (1)$$

$$NaHSe/Te + MX \longrightarrow MSe/Te + NaCl + HCl \quad\quad (2)$$

Where M= Cd, Pb and X = Cl_2, CO_3^{2-}, CH_3COO^- or NO_3^{2-}

Scheme 1. Equations for the formation of cadmium and lead selenide/telluride nanoparticles

CdSe NANOPARTICLES
The absorption spectra for both the HDA and TOPO capped CdSe nanoparticles show characteristic excitonic features (Figure 1).The absorption band edge as calculated using the direct band gap method[26] are 2.09 eV (593 nm, HDA) and 2.03 eV (618 nm, TOPO), both blue-shifted in relation to bulk CdSe, 1.73 eV (716nm). The UV spectra also show distinct excitonic features at 2.20 eV (563nm, HDA) and 2.15 eV (576nm, TOPO) which can be attributed to the first electronic transition (1s-1s) occurring in CdSe nanoparticles.[27]

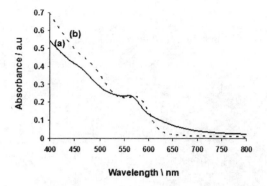

Figure 1. Absorption spectra of (a) HDA and (b) TOPO capped CdSe synthesized using CdCl$_2$ as the cadmium source.

The photoluminescence spectra for both the HDA and TOPO capped particles exhibit band-edge luminescence for excitation at 400 nm (Figure 2). The emission peaks are narrow confirming the monodispersity of the particles and the emission maxima show a red shift in relation to the corresponding absorption band-edge. The narrow emission line width indicates the growth of the crystallites with few electronic defect sites.

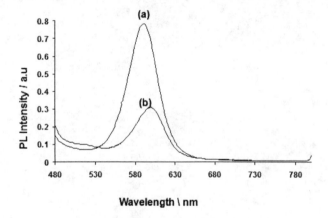

Figure 2. Photoluminescence spectra (λ_{400nm}) of (a) HDA and (b) TOPO capped CdSe

The TEM images (Figures 3a and b) of the as-prepared CdSe particles show well-defined, monodispersed, spherical particles. The average particle sizes as calculated from the TEM images are 3.0 ± 0.3 nm and 3.8 ± 0.4 nm for the respective HDA and TOPO capped CdSe particles. However there is a distinct change in particle shape when CdCO$_3$ is used as the cadmium source. HDA capped CdSe nanoparticles in the form of short and long rods are present, with average lengths of 20.74 ± 3.10 nm and 35.05 ± 3.19 nm respectively (Figure 3 c). In the case of TOPO capped particles, rod shaped particles with an average length of 35.34 ± 6.17 nm and breadth of 10.79 ± 1.58 nm are visible in the TEM image (Figure 3d). This trend is also observed for CdTe and PbTe nanoparticles and will be discussed in more detail in the latter part of this paper.

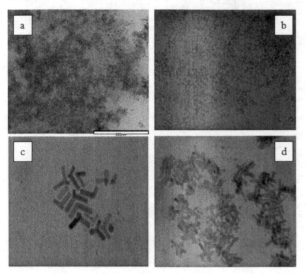

Figure 3. TEM images of CdSe nanoparticles from CdCl$_2$ (a) HDA capped (b) TOPO capped and from CdCO$_3$ (c) HDA capped and (d) TOPO capped.

CdTe NANOPARTICLES

The UV-Vis absorption spectrum (Figure 1) of the HDA-capped CdTe nanoparticles prepared from CdCl$_2$ exhibited a band gap at 691 nm (Figure 4), a blue shift in relation to bulk CdTe (795 nm). The distinct excitonic shoulder characteristic of monodispersed CdTe nanoparticles is visible at *ca.* 620 nm. The corresponding photoluminescence spectrum shows defect free narrow band-edge emission, with the emission maximum at 668 nm.

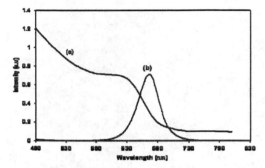

Figure 4. HDA-capped CdTe nanoparticles synthesized using $CdCl_2$ as the cadmium source (a) UV-Vis absorption and (b) photoluminescence (PL) spectra

The TEM image of the HDA capped CdTe nanoparticles synthesized showed monodispersed close to spherical particles with an average size of 5.51 nm ± 11 % (Figure 5a). The relatively narrow size distribution from the TEM measurements is consistent with the sharp band edge observed in the absorption spectrum. There is some aggregation of particles within the capping group matrix. The high resolution TEM image shows a well defined single particle with a diameter of 5.3 nm (Figure 5b). We decided to vary the cadmium source Figure 5c shows HDA capped CdTe nanoparticles synthesized using cadmium acetate (CH_3COOH) as the cadmium source at a reaction temperature of 230 °C. Particles with a slightly irregular shape showing some degree of ordered aggregation are observed.. The particles tend to form chains with attachment at the ends of the quasi spherical particles (see arrows). The epitaxial fusion of particles through aggregation whereby the lattice planes are aligned relative to each other is known as oriented attachment. This growth mechanism has been reported for several systems such as CdTe, CdSe, ZnO, CdS and PbSe nanorods. The process is driven by the reduction in surface area that occurs during the aggregation process.[28]

The HRTEM image (Figure 5d) shows particles imperfectly aligned, kinks or disruptions in the lattice planes in the adjacent particles are visible (shown by arrows). Further evidence for oriented attachment growth mechanism is observed for the CdTe particles synthesized from cadmium nitrate at 230 °C. Well defined nanorods are observed in the TEM image (Figure 5e). The HRTEM shows a single rod shaped particle with discontinuities in the lattice fringes which are due to dislocations and stacking faults, an indication that the rods are not perfect single crystals.

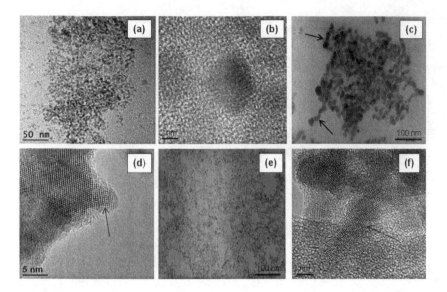

Figure 5. TEM and corresponding HRTEM images of HDA capped CdTe synthesized at 230 °C using (a,b) $CdCl_2$ (c,d) $Cd(CH_3COO)_2$ and (e,f) $Cd(NO_3)_2$

Rod shaped particles are also observed when cadmium carbonate is used at a similar temperature. (Figure 6a) The HRTEM image shows a well defined rod shaped particle (Figure 6b). The lattice planes show no discontinuities with the lattice spacing. The XRD studies for all samples reveal the existence of the cubic zinc blende phase of CdTe.

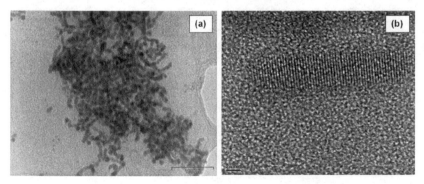

Figure 6. (a) TEM and corresponding (b) HRTEM images of HDA capped CdTe synthesized at 230 °C using $CdCO_3$.

PbSe NANOPARTICLES

Figures 7a and c shows the TEM images of PbSe particles prepared using PbCO₃ at reaction temperatures of 190 and 230 °C. There was a remarkable change in shape of nanoparticles with temperature observed. At 190 °C (Figure7a), monodispersed rods of PbSe nanoparticles were observed with an average width of 12 ± 2 nm and average length of 30 ± 2 nm. The HRTEM image (Figure 7b) shows lattice fringes with the d-spacing 3.05 Å corresponding to the (100) reflection of cubic PbSe. However at 230 °C, close to perfect cubes were observed with an average size of 15 ± 2 nm (Figure 7c). The factors controlling the shapes of inorganic nanocrystals involve the competition between thermodynamic and kinetic factors.[29] The balance between kinetic and thermodynamic factors will determine the growth process and final morphology of the particles. High temperatures favour the stable thermodynamic control resulting in the formation of cubes. Lower temperatures facilitate the less stable kinetic growth process leading to the formation of rods.

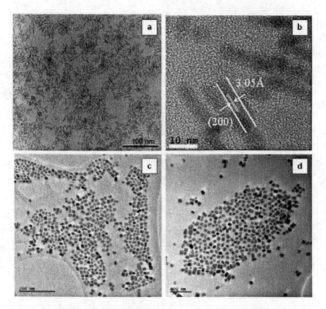

Figure 7. TEM image of HDA capped PbSe synthesized from PbCO₃ at (a) 190 °C (b) corresponding HRTEM image showing rods, (c) at 230 °C and (d) from Pb(NO₃)₂ at 230 °C

Figure 7 shows TEM images of PbSe nanoparticles obtained using Pb(NO₃)₂ as the lead source at a reaction temperature of 230 °C (Figure 7d). Particles in the shape of cubes similar to those obtained using the carbonate source at the same temperature are observed. The average size of the cubes is 19 ± 1.5 nm.

The PbSe nanoparticles using $PbCO_3$ were examined by X-ray powder diffraction to analyse the crystal structures and phase compositions. All the XRD patterns could be assigned to the face centered cubic phase of PbSe with lattice constant of $a = 6.124$ Å. The major diffraction peaks were the (111), (200), (220) and (311) planes of cubic PbSe.

PbTe NANOPARTICLES

PbTe nanoparticles were prepared by the same method using tellurium instead of selenium. In experiments using the lead carbonate, PbTe nanoparticles obtained at two reaction temperatures are shown in Figure 7. At 230 °C nanorods with lengths of 35 ± 3 and widths of 8 ±1nm were observed (Figure 7a). The dimensions of the rods increased with a increase in the reaction temperature to 270 °C, whereby rods with the length of 51.7 ± 5 nm and width of 9.1 ± 1.5 nm were observed.

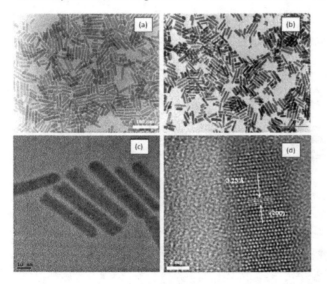

Figure 7. TEM images of PbTe using $PbCO_3$ at (a) 230 °C and (b) 270 °C, (c,d) HRTEM images of the sample at 270 °C.

The HRTEM images of the PbTe nanorods synthesized at 270 °C show clearly visible lattice fringes with a lattice spacing of 3.23 Å, assigned to the (200) reflection of cubic PbTe (Figure 7 c and d). The XRD pattern indicates the predominance of the halite (fcc, space group Fm3m) phase with the lattice constant of $a = 6.449$ Å which is consistent with the standard value ($a = 6.454$ Å) of bulk face centered cubic phase of PbTe. The major diffraction peaks are indexed as (200), (220), (222), (420) and (422) planes of cubic PbTe.

The PbTe nanoparticles were also prepared using $Pb(NO_3)_2$ and $PbCl_2$ as the lead sources at a reaction temperature of 230 °C. . Both the lead nitrate and chloride gave mostly spherical particles. The average particle size of the nitrate sample was 12.8 ±1.5 nm (Figure 8a) whereas the chloride sample gave spheres with an average diameter of 12.5 ±2 nm (Figure 8b) There were no evidence of anisotropic particle growth with both the lead nitrate and chloride sources.

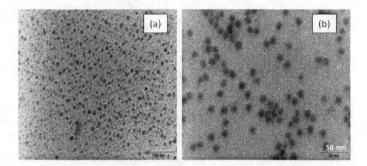

Figure 8. TEM images of PbTe nanoparticles synthesised at 230 °C from (a) $Pb(NO_3)_2$ and $PbCl_2$.

EFFECT OF METAL SOURCE ON PARTICLE MORPHOLOGY

In the case of all the cadmium and lead selenide and telluride nanoparticles the metal source has an influence on the final morphology of the particles. In CdSe, both the TOPO capped and HDA capped nanoparticles had a rod shaped morphology when cadmium carbonate was used as the metal source. In the case of CdTe we propose two mechanisms of growth. For the nanorods of CdTe obtained using the nitrate and acetate cadmium salt there is evidence that growth occurs via an oriented attachment mechanism.[28] The TEM image of the CdTe particles synthesized using the acetate salt shows particles fused together to form a chain. (Figure 5c). The HRTEM of both the CdTe obtained from the acetate and nitrate source show discontinuities or disruptions in the lattice planes, further evidence of growth via an oriented attachment process. However the growth mechanism of the CdTe, CdSe, PbTe and PbSe nanoparticles synthesized from the carbonate metal salt occurs via the traditional Ostwald ripening process. The anisotropic nature of the particles could be attributed to the lower solubility of the metal carbonate. This results in a suspension of the lead carbonate thereby forming a shell on the nanoparticle surface. The surface energy of the crystallographic faces of the initial formed seeds have a dominant effect on the anisotropic growth pattern of the nanoparticles. The surfactant, in this case hexadecylamine can be selectively adsorbed onto the surface thereby modulating the surface energy. The undissolved cadmium carbonate also selectively adheres to the surface of the nanoparticles thereby accentuating the difference in the growth rate between the crystallographic phases. The result is anisotropic growth in the form of nanorods.

CONCLUSIONS

High quality CdSe, CdTe, PbSe and PbTe nanocrystals have been prepared by a simple hybrid solution based high temperature route. The route involves the reduction of the chalcogenide in solution followed by the addition of the metal salt and thermolyis in a coordinating solvent. The route does not use any highly toxic or noxious starting materials and the reaction conditions are relatively mild thereby rendering it environmentally friendly.

The ability to control the shape of the particles by variation of the metal salt also makes it a reliable and easy route to anisotropic particles than earlier approaches.

ACKNOWLEDGEMENTS
The author acknowledges the Department of Science and Technology (DST) and National Research Foundation (NRF) of South Africa through the DST/NRF South African Research Chairs Initiative (SARCHi) for funding.

REFERENCES
1.A.P. Alivisatos, Semiconductor Clusters, Nanocrystals, and Quantum Dots, *Science*, 271, 933- 37, (1996).
2. N.P. Gaponik, D.V. Talapin, A.L. Rogach and A. Eychmuller, Electrochemical Synthesis of CdTe Nanocrystal/Polypyrrole Composites for Optoelectronic Applications, *J. Mater. Chem.*, 10, 2163-66, (2000).
3. W.U. Huynh, J.J. Dittmer and A.P. Alivisatos, Hybrid Nanorod-Polymer Solar Cells, *Science*, 295, 2425-27, (2002).
4. J. Lee, A. O. Govorov and N. A. Kotov, Impact of noble metal nanostructures on surface trapping state of semiconductor quantum dots, *Nano Lett.*, 4, 2323-26, (2004).
5. S. A. Mc Donald, G. Konstantatos, S. Zhang, P. W. Cyr, E. J. D. Klem, L. Levina and E. H. Sargent, Solution-processed PbS quantum dot infrared photodetectors and photovoltaics, *Nat. Mater.*, 4, 138-142, (2005).
6. C.B. Murray, D. J Norris and M.G. Bawendi, Synthesis and characterization of nearly monodisperse CdE (E = sulfur, selenium, tellurium) semiconductor nanocrystallites, *J. Am. Chem. Soc.*, 115, 8706-15, (1993).
7. M. A. Malik, N. Revaprasadu and P. O' Brien, Air-Stable Single Source Precursors for the 'One-pot' Synthesis of Chalcogenide Semiconductor Nanopaticles. *Chem. Mater.*, 13, 913-20, (2001).
8. Z.A. Peng and X. Peng, Formation of high-quality CdTe, CdSe, and CdS nanocrystals using CdO as precursor, *J. Am. Chem. Soc.*, 123, 183-4, (2001).
9. X. Peng, L. Manna, W.D. Yang, J. Wickham, E. Scher, A. Kandavich and A.P. Alivisatos, Shape Control of CdSe Nanocrystals, *Nature*, 404, 59-61, (2000).
10. D.V. Talapin, S. Haubold, A.L. Rogach, A. Kornowski, M. Haase and H. Weller, A novel organometallic synthesis of highly luminescent CdTe nanocrystals, *J. Phys. Chem. B*, 105, 2260-63,(2001).
11. L. Zou, Z. Gu, N. Zhang, Y. Zhang, Z. Fang, W. Zhu and X. Zhong, Ultrafast synthesis of highly luminescent green- to near infrared-emitting CdTe nanocrystals in aqueous phase, *J. Mater. Chem.* 18 2807-15, (2008).
12. M. Green, H. Harwood, C. Barrowman, P. Rahman, A. Eggeman, F. Festry, P. Dobson, and T. Ng, A facile route to CdTe nanoparticles and their use in bio-labelling, *J. Mater. Chem.* 17, 1989-94, (2007).
13. C. B. Murray, S. Sun, W. Gaschler, H. Doyle, T. A. Betley and C.R. Kagan, Colloidal synthesis of nanocrystals and nanocrystal superlattices, *IBMJ. Res. Dev.* 2001, 45-56, 47.
14. W. Lu, J. Fang, Y. Ding and Z. L. Wang, Formation of PbSe nanocrystals: A growth towards nanocubes, *J. Phys. Chem. B*, 109, 19219-22, (2005).
15. K. Cho, D. V. Talapin, W. Gaschler and C. B. Murray, Designing PbSe nanowires and nanorings through oriented attachment of nanoparticles, *J. Am. Chem. Soc.*,127,7140-47,(2005).
16. W. Z. Wang, Y. Geng, Y. T. Qian, M. R. Ji, X. M. Liu, Photochemical synthesis and characterization of PbSe nanoparticles, *Adv. Mater.* 10, 1479, (1998).

17. S. M. Lee, Y. Jun, S. N. Cho, J. Cheon, Single-crytsalline star shaped nanocrystals and their evolution: Programming the geometry of nano-building blocks, *J. Am. Chem. Soc.,* **124**, 11244-45, (2002).
18. M. Du, Y. Wang, J. Xu, P. Yang and Y. Du, PbSe quantum dots: Preparation in a high boiling point solvent and characterisation, *Colloid. J.* **70**, 720, (2008).
19. X. Chen, X, T.J. Zhu, X.B Zhao, Synthesis and growth mechanism of rough PbTe nanorods polycrystalline tehrmoelectric nanorods, *J. Cryst. Growth,* **311**, 3179-83, (2009).

20. S.O Oluwafemi and N. Revaprasadu, A new synthetic route to organically capped cadium selenide nanoparticles, *New J. Chem.* **10**, 1432-37, (2008).
21. N.N Maseko, N. Revaprasadu, V.S.R. Rajasekhar Pullabhotla, K. Ramasamy and P.O' Brien, The Influence of the Cadmium Source on the Shape of CdSe Nanoparticles, *Materials letters,* 64, 1037-9, (2010).
22. N. Ziqubu, K. Ramasamy, N. Revaprasadu, V.S.R. Rajasekhar Pullabhotla and P. O'Brien A Simple to Route to Dots and Rods of PbTe Nanoparticles, *Chemistry of Materials,* **22**, 3817-19, (2010).
23. N.M Mntungwa, V.S.R Rajasekhar Pullabhotla and N. Revaprasadu, A Facile Route to Shape Controlled CdTe nanoparticles, *J. Mater. Chem and Phys,* 126, 500-6, (2011).
24. K. Ramasamy, N. Ziqubu, V. S. R. Rajasekhar Pullabhotla. O.A. Nejo, A A. Nejo, N. Revaprasadu, and P. O'Brien, A New Route to Lead Chalcogenide Nanocrystals *Eur. J. Inorg. Chem.,* 5196-5201, (2011).
25. U. K. Gautam, M. Nath, and C. N. R. Rao, New strategies for the synthesis of *t*-selenium nanorods and nanowires, *J. Mater. Chem.,* **13**, 2845-47, (2003).
26. J.L. Pankove, *Optical Processes in Semiconductors*, Dover Publications, New York. 1990.
27. J. E. Bowen-Katari, V. L. Colvin, and A. P. Alivisatos, X-ray photoelectron spectroscopy of CdSe nanocrystals with applications to studies of the nanocrystal surface. *J. Phys. Chem.,* 98, 4109-17, (1994).
28. R. L. Penn, J. F. Banfield, Imperfect orientated attachment: Dislocation in defect-free nanocrystals, *Science,* 281, 969-71, (1998).
29. S.M. Lee, S.N. Cho, J. Cheon, Abnisotropic shape control of colloidal inorganic nanocrystals, *Adv. Mater.,* **15**, 441, (2003).

NANOWIRE BASED SOLAR CELL ON MULTILAYER TRANSPARENT CONDUCTING FILMS

D. R. Sahu[1], Jow-Lay Huang[2] and S. Mathur[3]

1. School of Physics, Materials Physics Research Institute and DST/NRF Centre of Excellence in Strong Materials, University of the Witwatersrand, Private Bag 3, Wits 2050, Johannesburg, South Africa
2. Department of Materials Science and Engineering, Center for Micro/Nano Science and Technology, Research Center for Energy Technology and Strategy, National Cheng Kung University, Tainan 701, Taiwan
3. Institute of Inorganic Chemistry, University of Cologne, Greinstraße 6, D-50939 Cologne, Germany

ABSTRACT

Multilayer transparent conductive oxide (TCO) thin films of three alternative layers ZnO/Ag/ZnO (named as ZAZ) have been used for the fabrication of nanowire based solar cells. Multilayer transparent conductive ZAZ films were grown using RF and dc sputtering method for deposition of ZnO and Ag layers simultaneouly. Nanowire ZnO was grown on this TCO using chemical bath deposition (CBD). These nanowire solar cells on the ZAZ coating yielded an overall cell efficiency of 3.41% at one sun light intensity. The dye sensitization process with the low cost mercurochrome is sensitive in this case of ZnO based multilayer.

1. INTRODUCTION

Solar cells are attractive candidates for clean and renewable power. With miniaturization, they might also serve as integrated power sources for nanoelectronic systems. One dimensional (1D) nanostructures such as nanorods and nanowires can be used to improve collection rate of carrier charges and energy efficiency of solar cells, to demonstrate carrier multiplication and to enable low-temperature processing of photovoltaic devices [1-3]. The unique properties of 1D nanostructures stem from their large surface to volume ratio, very high aspect ratio, and carrier and photon confinement in two dimensions. The high surface area of 1D nanostructures and the ability to synthesize radial heterostructures results in enhancing cell efficiencies by facilitating photon absorption, electron transport and electron collection. In the technical aspects, solar cells based on hybrid nanoarchitectures can generally suffer from relatively low efficiencies and poor stabilities due to the uncontrollable trapping sites such as oxygen defects, metallic catalysts and impurities [4-7]. Therefore, there is a need for the fabrication of novel nanowire-based solar cells using superior transparent electrode and active anode materials via structure and surface modification of metal oxide nanowires.

Zinc oxide, which has the wide band gap with good carrier mobility, is an attractive potential candidate for solar cell anode. Electron mobility in ZnO is much higher than in TiO_2, while the conduction band edge of both materials is located at the same level approximately [8]. A large range of fabrication procedures like Chemical vapor deposition [9, 10], sol-gel processes [11], chemical bath deposition [12,13] and vapor deposition [14] etc. are available for growth of ZnO nanostructures. In addition, ZnO possesses the preferred growth direction of c-axis, making it possibly to be grown on the substrate vertically and to provide a fast electron transport route for solar cells. The improvement of the electron transport in the anode of the dye sensitized solar cells (DSSCs) has been reported by using the single crystal and vertical ZnO nanowire (NW) arrays on transparent conductive oxides (TCOs) [15.16]. However, efficiencies of those solar cells are very low in compared to TiO_2 based DSSC. However, there is a significant improvement of the efficiency of the ZnO nanowire dye-sensitized

45

solar cell by the chemical bath deposition of the dense nanoparticles (NP) within the interstices of the vertical ZnO-NW anode [17]. There is also considerable enhancement of the efficiency of the ZnO-NW array/NP composite DSSC in comparison with the ZnO-NW [18]. On the other hand, mercurochrome ($C_{20}H_8Br_2HgNa_2O_6$) is one of the best photosensitizer for ZnO photoanode [19] to date and is much cheaper than the Ru complex dyes. Therefore, there is an upfront research for the enhancement of efficiency of NW based solar cells by the combination of process development and use of superior materials architecture. In this present work, mercurochrome sensitized ZnO-NW solar cells were fabricated on the developed ZnO based multilayer TCO to study the performances of the ZnO-NW solar cells.

2. EXPERIMENTAL

2.1 Synthesis of TCO
The thin films of multilayer ZnO/Ag/ZnO (ZAZ) structures were sputter deposited on glass (corning 1737F) using a zinc oxide (99.9995 purity, 7.62 cm diameter, 0.64 cm thickness, target materials Inc.) and metal Ag targets (99.999% purity, 7.62 cm diameter, 0.64 cm thickness, target materials Inc. in an inline magnetron sputter deposition system. The glass substrate was ultrasonically cleaned in acetone, rinsed in deionized water and subsequently dried in flowing nitrogen gas before deposition. Details of the deposition procedure were described in reference [20]. This developed TCO was used for the growth of nanowires along with nanoparticles. Fig. 1 shows the configuration of the multilayer ZAZ films. The thicknesses of the glass attached ZnO was kept 20 nm. The thickness of the Ag layer in case of ZnO multilayer was kept 6 nm according to our previous study [20]. The thickness of the outer surface layer was varied from 0 to 50 nm. The thickness of each multilayer was controlled by the deposition time. The sheet resistance of each transparent film was determined using four-point probe method. Transmittance spectra were obtained to study the optical properties of the transparent film.

2.2 Synthesis and characterization of ZnO-NWs anodes
High-density ZnO nanowires were grown on the developed TCO substrates (6 Ω per-square) by CBD method [12]. ZnO NWs were grown on the surface ZnO layer which acts like seed layer for growth of nanowires using CBD method in a 0.02 M aqueous solution of zinc acetate and hexamethylenetetramine ($C_6H_{12}N_4$, HMTA, Riedel-de Haen, 99.5%) at 95^0C for 3h. Multi-bath process was used for lengthening the NWs. Finally, as-grown ZnO NWs were annealed again in air at 400^0C for 30 min to improve the crystallinity of the NWs and the interfacial structures. NPs were synthesized among the ZnO NWs using a base-free CBD method. The annealed ZnO-NW sample was immersed in a 0.15M methanolic solution of zinc acetate dehydrates at 60^0C for 18h. The composite film was rinsed by methanol and dried at room temperature. The morphologies of the deposited materials were examined using fieldemission scanning electron microscopy (FESEM, JEOL JSM-7000F). The crystal structures of the composite films were investigated using transmission electron microscopy (TEM, JEOL 2100F)

2.3 Dye adsorption and fabrication of the ZnO NW/NP composite solar cells
Mercurochrome dye adsorption were carried out by immersing ZnO NW anodes into the $6.6 \times 10^{-4}M$ anhydrous ethanol solution of mercurochrome ($C_{20}H_8Br_2HgNa_2O_6$, Acros) at 50^0C for 8h. Sensitized NW films were rinsed with anhydrous ethanol to remove unadsorbed dye molecules on the surface of ZnO NW. The ZnO NW anodes were sandwiched together to the platonized counter electrodes by hot melt spacers (SX 1170, 25 μm thickness, Solaronix). The space of cells were filled with the liquid electrolyte composed of 0.3 M tetra-n-propylammonium iodide, Pr₄NI

((CH$_3$CH$_2$CH$_2$)4NI, 98%, Avocado) and 0.03 M iodine (I2, Showa) in an ethylene carbonate and acetonitrile mixed solvent (60:40 by volume) as well as or 0.1M LiI (ProChem), 50 mM I2, 0.5 M

Dye adsorption of the ZnO-NW array/NP composite anodes were conducted in a refluxing 1.6 × 10^{-4} M dry methanolic solution of mercurochrome at 65°C for 20 min then rinsed by anhydrous ethanol. The sensitized NW/NP composite electrode and plantinized FTO counter electrodes were sandwiched together with a 25 μm-thick hot-melt Surlyn spacers (SX 1170-25, Solaronix SA). The liquid electrolyte solution was introduced between the sensitized and counter electrodes by capillary action. The cells were then sealed by AB paste. A black-painted mask was used to create an exposed area of 0.04 cm^2 for all samples. Photocurrent (J) and photovoltage (V) were measured using Keithley 2400 Source meter under AM 1.5 simulated sunlight at 100 mW cm^{-2} (300 W, model 91160A, Oriel). Typically from current (J)-voltage (V) characteristics of the solar cell, one can calculate the efficiency of the solar cell. The x-axis stands for the applied voltage and the y-axis shows the current density per square centimeter. Here, we define the current when the voltage is zero as the "short-circuit current density, Jsc" and the voltage when the current is zero is called the "open-circuit voltage, Voc". If we multiply each voltage and its responding current in the curve, then compared all of the values, a maximum power (P$_{max}$) will be gained. The efficiency of the cell then can be expressed as,

$\eta = J_{sc} \times V_{oc} \times F.F / P_{inc}$

where the η, P_{inc} are efficiency and intensity of the incident light. The FF is the fill factor which is expressed as F. $F = P_{max} / J_{sc} \times V_{oc}$.

3. RESULTS AND DISCUSSION

3.1 Characterization of TCO substrate

Fig. 2 shows the transmission spectra of multilayer coatings (thickness of the surface ZnO was varied from (0 to 50 nm), ITO and FTO. In Fig. 2 (a) the transmittance of multilayer with a ZnO surface layer thickness of 50 nm was comparable with those of FTO and ITO which suggests together with the good conductivity (6 Ω/Sq.) and transmittance above 90 %, that the multilayer be useful for application to solar cell as a substitute for FTO and ITO. In addition to this other multilayer having better properties are also placed for comparison. The optimum values of the transmittance and reflectance occur for coating with 20 nm of both surface ZnO. The transmittance and reflectance deteriorate at the surface layer thicknesses of more than 50 nm because of interferences in the surface ZnO layer. The multilayer with the thickness of 20 nm/ 6 nm/20 nm was found to show the optimum transmittance of 0.91 in the visible range. The cross-section view of the multilayer film ZAZ on glass substrate confirm that each layer of ZnO and Ag has flat and smooth structure which suggests high conductivity at the Ag layer of ZAZ.

3.2 Characterization of NWs and NW array

Fig. 3a and b show the top view and cross-sectional SEM images of aligned ZnO NWs grown on the TCO substrate after the aqueous CBD growth. The ZnO NW array have diameter in the range of 20–350 nm. Fig. 3c shows the high-resolution (HR) TEM image of an individual NW, illustrating that the NW possesses a single crystal structure and lattice spacing of around 0.52 nm along the longitudinal axis direction which corresponds to the $d$spacing of ZnO (001) crystal planes. Fig.4 shows the top view and cross-sectional SEM images of the NW-NP films formed by growing NPs within the interstices of ZnO NW arrays. Using longer growth period for NP, more particles are grown within the interstices of the NW array forming a denser NW-NP composite film [21].

3.2.1 The performances of the N3- and mercurochrome-sensitized NW solar cell

The mercurochrome and N3-sensitized ZnO NW DSSCs are used to investigate the performances of the ZnO NW solar cell. The mercurochrome dye is proper for ZnO DSSC in comparison with Ru-complex dyes. The JV curves of the ZnO NW DSSCs are shown in the Fig. 5. The Jsc, Voc, F.F. and η of the N3- and mercurochrome-sensitized ZnO NW DSSCs are 3.42 and 3.11 mA/cm^2, 0.61 and 0.59 V, 0.52 and 0.61 as well as 1.08 and 1.12, respectively. The Jsc and Voc of N3/ ZnO NW DSSC are higher than those of mercurochrome-sensitized one. But the F.F. of the mercurochrome/ ZnO NW DSSC is larger than that of N3-sensitized one, leading to the higher efficiency of the mercurochrome-sensitized ZnO NW DSSC in comparison with the N3-sensitized DSSC. Due to a considerable higher fill factor, mercurochrome-sensitized ZnO NW DSSC shows higher efficiency than N3-sensitized one. Since the ZnO-NW arrays on TCO substrates employed in both cells are fabricated using the same procedure, all resistances in the both photoanodes are suggested to be the same. The lower F.F. in the N3/ZnO-NW DSSC implies that the photoelectrons in the N3-sensitized electrode are easier to back react with I3- in the electrolyte in comparison with the mercurochrome-sensitized one[12]. Similar type of behavior was also observed using ZnO nanowires grown on FTO substrates by Wu et al [12]. Generally, the nanowire shows nearly the linearly increase in Jsc an the nanowire films has the high collection efficiency. The NW cells also generate higher currents compared to the ZnO particle cells, showing the nanowire photoanode as a superiority charge collector. Law et al. has reported dye sensitized solar cells (DSSC) with single-crystalline ZnO NW arrays vertical to TCO has overall efficiency (η) of 1.5 %, still inferior to those of the TiO$_2$ NP cell (N719 dye) [15]. They suggested that the performance to the lower roughness factor of ZnO NW. Apart from the low photocurrent, the low FF in the ZnO-NW DSSC, small shunt resistance or/and severe electron interfacial recombination is also a restriction for its efficiency [12].

3.2.2 The performances of the ZnO NW/NP composite
It is reported that the well-aligned ZnO NW/NP composite anode can enhance the performance of the ZnO NW DSSC [22, 22]. This composite structure using nanoparticles provide a large surface for dye adsorption and the nanowires contribute to fast routes for electron transport. Therefore, the ZnO NWs and the ZnO NWs/NPs composite film were used for the assembly of the nanowire dye sensitized solar cells (DSSCs). The J –V curves of the mercurochrome-sensitized ZnO NW-array/ NP composite DSSC is shown in Fig. 6. The current density, Voc and F.F. of the ZnO NW DSSC are enhanced by filling NPs into the ZnO-NW array. The higher Jsc of the ZnO NW-array/ NP composite DSSC compared to that in the ZnO NW DSSC is due to the NPs provide larger surface for dye adsorption. Since the same electrolyte are employed in both ZnO DSSCs, the higher Voc of the ZnO NW/NP composite DSSC is contributed to the lower recombination and higher amounts of injected electrons [21]. Maximum efficiency of 3.41% is achieved using a mercurochrome-sensitized composite film which is composed of ZnO NPs and well-aligned NWs. Heterogeneous nucleation of the NPs growth on the surface of the ZnO nanowires using CBD route is crucial for the formation of the dense NW/NP composite films. A considerable efficiency enhancement is observed in NW/NP DSSC compared to the ZnO NW one. The origin of the improve efficiency is suggested to the nanoparticles in the ZnO composite anode increase the surface area for dye adsorption and reduce of the electron back reaction on the exposed TCO surface. The improvement in the conversion efficiency also depends not on the differences in the conductivities of the transparent layers but on the antireflection of the multilayers and the transmittance of the incident light. It is known that the introduction of anti reflection films on solar cells can increase the photovoltaic performance of the cells. This proper arrangement of the thickness of each layer in the ZAZ and ZnO nanostructure suppress internal reflection achieves a high photocurrent and conversion efficiency of the DSSC. Higher dye absorbance on the ZnO/ZAZ structure also indicates higher free electron density in the conduction band which enables higher Jsc. Therefore, fast electron transport in the transport medium

and the efficient electron collection of the transparent electrode are of critical importance to improve the conversion efficiency of solar cells. Further, if the dye specifically designed for ZnO is available, the efficiency of the ZnO NW/NP DSSC higher than that of TiO_2 NP one can also be achievable.

4. CONCLUSION

The developed TCO using sputtering with resistance of 7 Ω/sq and average transmittance of 85-90 % were found suitable as transparent electrode for solar cells. The mercurochrome- and N3-sensitized ZnO NW and ZnO NW/NPS DSSCs fabricated on developed TCO indicates superior performance than the ITO or FTO based solar cells. ZnO NW/NP composite structure shows better performance than the NW solar cell with mercurochrome sensitizer. . The solar cells on the ZAZ coating yielded an overall cell efficiency of 3.41 % at one sun light intensity.

ACKNOWLEDGEMENT:
Authors are thankful to the students of Prof. J. J. Wu for their useful help in analysis and growth of nanowires on multilayer TCO. Financial support received from National research Function (NRF), South Africa and BMBF, Germany for the SA/Germany research cooperation programme is highly acknowledged.

REFERENCES:
[1] M. Law, L. E. Greene, J. C. Johnson, R. Saykally, P. Yang, Nat. Mater. 4, 455, 2005.

[2] Y.J. Lee, D. S. Ruby, D. W. Peters, B. B. McKenzie, J. W. P. Hsu, Nano Lett. 8, 1501, 2008

[3] S. M. Mahpeykar, J. Koohsorkhi, H. Ghafoori-Fard, Nanotechnology, 23, 165602, 2012

[4] A. C.Fisher, L. M. Peter, E. A. Ponomarev, A. B. Walker, K. G. U Wijayantha, J. Phys. Chem. *B* 104, 949, 2000.

[5] Q. Zhang, S. Yodyingyong, J. T, D. G. Z. Cao, Nanoscale 4, 1436, 2012

[6] K. D. Benkstein, J. van de Lagemaat, N. Kopidakis, A. J. Frank, J. Phys. Chem. B 107, 7759, 2003

[7] Z. Qin, Y. H. Huang, J. J,. Qi, Q. L. Liao, W. H. Wang, Y. Zhang, Mater. Lett. 65, 3506, 2011

[8] Frank,S. N.; Bard, A. J. *J. Phys. Chem.* 81, 1484, 1997

[9] L. Y. Wen, K. M. Wong, Y. G. Fang, M. H. Wu, Y. Lei, J. Mater. Chem., 21, 7090, 2011

[10] J. H. Zeng, Y. L. Yu, Y. F. Wang, T. J. Lou, Acta Materilia, 57, 1813, 2009

[11] M. Guo, P. Diao, S. Cai, J. *Solid State Chem.* 178, 1864, 2005

[12] J. J. Wu, G. R. Chen, H. H. Yang, C. H. Ku, J. Y. Lai, *Appl. Phys. Lett.* 90, 213109, 2007

[13] D. Lincot, MRS Bull. 35, 778, 2010

[14] J. J. Wu, S. C. Liu, *Adv. Mater.* 14, 215, 2002

[15] M. Law, L. E. Greene, J. C. Johnson, R. Saykally, and P. Yang, Nat. Mater. 4, 455, 2005.

[16] J. B. Baxter and E. S. Aydil, Appl. Phys. Lett. 86, 053114, 2004

[17] J. B. Baxter, A. M. Walker, K. V. Ommering, and E. S. Aydil, Nanotechnology 17, S304, 2006

[18] C. H. Ku, J. J. Wu, Nanotechnology, 18, 505706, 2007

[19] K. Hara, T. Horiguchi, T. Kinoshita, K. Sayama, H. Sugihara, and H. Arakawa, Sol. Energy Mater. Sol. Cells 64, 115, 2000

[20] D. R. Sahu, S. Y. Lin, J. L. Huang, Appl. Surf. Sci. 252, 7509, 2006

[21] C. H. Kuo, H. H. Yang, G. R. Chen, J. J. Wu, Cryst. Growth Design. 8, 283, 2008

[22] Z. G. Wang, M. Q. Wang, Z. H. Lin, Y. H. Xue, G. Huang, X. Yao, Appl. Surf. Sci, 255, 4705, 2009

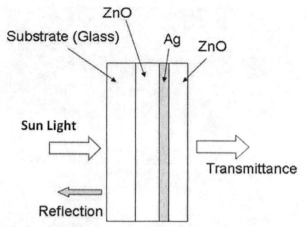

Fig. 1 Structure of transparent conducting ZAZ film

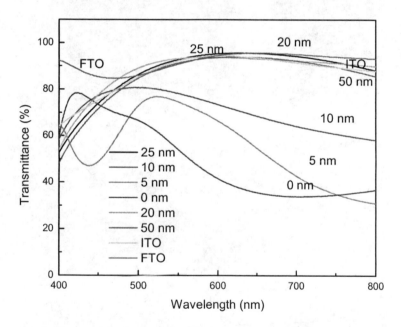

Fig. 2 The transmission spectra of multilayer coatings (a) thickness of the surface ZnO was varied from (0 to 50 nm), ITO, FTO

Fig. 3 (a) The top view and (b) cross-sectional SEM images of aligned ZnO NWs grown on the TCO substrate after the aqueous CBD growth.

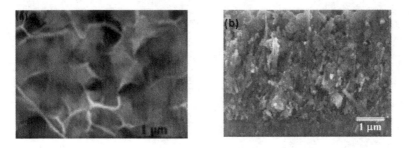

Fig. 4 The (a) top view and (b) cross-sectional SEM images of the NW-NP films formed by growing NPs within the interstices of ZnO NW arrays

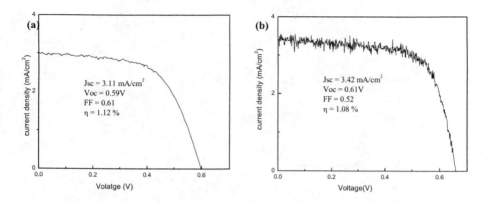

Fig. 5 The JV curves of the (a) mercurochrome-sensitized (b) N3 sensitized ZnO NW DSSCs

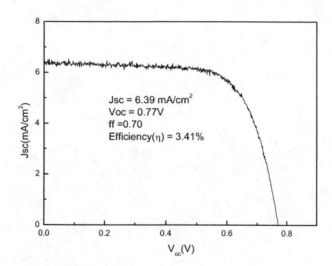

Fig. 6 The J –V curves of the mercurochrome-sensitized ZnO NW-array/ NP composite DSSC.

ANTIMICROBIAL PROPERTIES OF COPPER AND SILVER LOADED SILICA
NANOMATERIALS

Pavithra Maniprasad[γ], Roseline Menezes[γ], Jenelle Suarez[γ] and Swadeshmukul Santra[γ,μ,λ]

[γ]Burnett School of Biomedical Sciences, [μ]NanoScience Technology Center, [λ]Department of
Chemistry
University of Central Florida
Orlando, FL 32826

ABSTRACT
    Antimicrobial properties of copper (Cu) and silver (Ag) ions have been widely studied.
Hundreds of nanotech based consumer products are now available in the market which uses
antimicrobial Ag nanoparticles. Cu and Cu alloy based touch surfaces are shown to be effective in
controlling bacterial infection. In this study, we will present our research on synthesis and
characterization of sol-gel silica nanoparticle/nanogel materials loaded with antimicrobial Cu and Ag.
Structure/morphology and antimicrobial properties of the silica nanoparticle/nanogel delivery system
with and without containing the active agent (Cu or Ag) will be discussed. We have tested
antimicrobial properties of these materials against both gram-negative (E. Coli) and gram-positive (B.
Subtilis) bacteria. Our results on Cu nanomaterials showed improved antibacterial efficacy of Cu
loaded silica nanomaterial over its Cu source while the concentration of metallic Cu remained the
same. Several materials characterization techniques were used to understand structure-property
relationship using Cu loaded silica nanoparticle/nanogel nanomaterial.

INTRODUCTION
    Metallic nanoparticles have been used recently for a wide range of biomedical applications[1, 2].
Silver and copper have been known for ages for its antimicrobial properties. Copper and silver are
being used as antifouling, antifungal agents for many industrial applications. Their antibacterial
property finds wide usage in many health-care facilities to create microorganism free environment.
Various techniques are being used for synthesis of metallic nanoparticles. They include inert gas
condensation technique, electrolysis method, deposition of metallic salts on the matrix, reduction of
metallic salt[3-7]. This paper is focused on the synthesis, characterization and antibacterial properties of
silica-silver nanogel and silica-copper nanoparticles by acid and base hydrolysis respectively. Silver
embedded silica nanogel is synthesized using simple water based sol-gel technology. In this method
the simultaneous hydrolysis and condensation of silica facilitates the reduction of silver ions to form
silver nanoparticles entrapped in silica gel matrix[8]. Copper nanoparticles are synthesized based on a
novel core-shell design, where the silica nanoparticle serves as the 'core' and copper grows as the
'shell' around the core[9].

EXPERIMENTAL
    All reagents were purchased from commercial vendors and used without any further
purification. Bacterial strains were provided by Ishrath Sharma, Microbiology department, University
of Central Florida.

Nanomaterial synthesis
    The core-shell copper loaded silica nanoparticles (C-S CuSiNP) were synthesized in a two-step
fashion as discussed in our previous work[9]. The first step involved the synthesis of 'seed' silica
nanoparticles based on a published protocol[10]. Tetraethylorthosilicate (TEOS, a silane precursor;

Fisher Scientific) was added to a solution of 95% ethanol (Fisher Scientific), ammonium hydroxide (Fisher Scientific) and water (nanopure deionized) under stirring conditions. The contents were left on a 400 rpm magnetic stirrer for 1 hour. This was followed by sonication for 10 minutes. The silica nanoparticles (SiNP) were purified by washing (centrifuging) the particles with 95% ethanol at 10,000 rpm for 10 minutes to remove ammonium hydroxide. The final wash (10,000 rpm, 10 minutes) was done with water. The second step involved the growth of copper shell around the silica nanoparticle core. The silica 'seed' particles were dispersed in acidic pH water followed by addition of copper sulfate pentahydrate (CQ concepts, Ringwood, IL) and TEOS under magnetic stirring conditions. The growth of the copper 'shell' around the silica nanoparticle 'core' was allowed to grow for 24 hours. The particles were then isolated by washing (centrifugation) with water twice at 10,000 rpm for 10 minutes to remove excess copper.

The synthesis of silver loaded silica nanogel (AgSiNG) was carried on using sol-gel method in water and acidic condition in one step[11]. This procedure was similar to the second step of C-S CuSiNP synthesis as described in the above section. Silver nitrate salt (Acros organics) was used as the source of Ag and 1% Nitric acid (Macron) solution was used for the acid catalyzed hydrolysis of TEOS. Silica nanogel (SiNG) was prepared similarly (without silver nitrate) which was used as control. No further purifications were made to AgSiNG and SiNG materials except that the pH was adjusted to 7.0 using dilute sodium hydroxide solution. A pale yellow coloration was observed for the AgSiNG material.

Nanomaterial characterization

Zeiss ULTRA-55 FEG Scanning electron microscopy (SEM) was done to estimate the particle size and morphology of C-S CuSiNP. Spin coating technique was used to prepare the SEM sample on a silicon wafer. The amount of copper in C-S CuSiNP was quantified by Atomic Absorption Spectroscopy (AAS, Perkin Elmer AAnalyst 400 AA spectrometer). The metallic silver content in AgSiNG was quantified from the amount of silver nitrate added in the formulation. The AgSiNG formulation was completely transparent and the material could not be centrifuged down from the solution even at 10,000 rpm. Therefore the AgSiNG formulation was used for antimicrobial studies without any further purification. High-resolution transmission electron microscopy (HRTEM, Technai) technique was used to analyze AgSiNG material for the formation of silver nanoparticles. Sample was prepared by placing a drop of AgSiNG on a carbon coated copper grid.

Antibacterial assays

Antibacterial properties of C-S CuSiNP and AgSiNG materials were evaluated against a gram positive *Bacillus subtilis* (*B.subtilis*, ATCC 9372) and a gram negative *Escherichia coli* (*E.coli*, ATCC 35218) organism. For all antibacterial assays, bacterial concentration of 10^5 cells/mL was considered. Copper sulfate with same metallic copper concentration was used as positive controls and silica nanoparticle (without Cu loading) was used as negative control. In case of AgSiNG, silver nitrate with similar metallic silver concentration was used as positive control and silica gel (without silver) was used as negative control.

(i)   Bacterial growth inhibition in LB broth using turbidity

Two sets of different concentrations of C-S CuSiNP were made in LB broth (0.49, 1.2, 2.4, 4.9, 7.2 and 9.8 ppm copper concentration) to a final volume of 10 mL. 250µL (10^5 cells/mL) of *E.coli* and *B.subtilis* were added to set 1 and 2 respectively. Silica nanoparticle was used as negative control and copper sulfate with equivalent amount of copper was used as positive control. All the tubes were incubated at 37°C on a shaker at 150rpm for 24 hours. Aliquots were taken after 24 hours to measure optical density at 600nm. The same was repeated for AgSiNG.

(ii)   Bac-Light assay for live/dead cell staining

A known concentration of C-S CuSiNP and AgSiNP with appropriate controls was incubated with *E.coli* and *B.subtilis* in LB broth (10^5 cells/mL) to determine cell viability using the BacLight bacterial viability kit L7012[12]. The samples were incubated for 4 hours at 37°C on 150rpm shaker. The samples were then centrifuged at 10,000 ×g for 10 minutes. The supernatant was discarded and resuspended in 0.85% saline and centrifuged again. The final pellet was resuspended in 0.85% saline. 3μL of the baclight dye mixture was added to all tubes and incubated at room temperature in dark for 15 minutes. 5μL of the bacterial suspension was trapped between a slide and coverslip and viewed under a fluorescence microscope. The dead and live cells are counted using the red and green filters respectively.

RESULTS AND DISCUSSION
Nanomaterial characterization
    The SEM images of the 'core' SiNP (an average particle size of ~ 380 nm) and the C-S CuSiNP (an average particle size of ~ 450 nm) are shown in **Figure 1a** and **Figure 1b**, respectively. The increase in particle size and spherical morphology for C-S CuSiNP confirms uniform growth of copper loaded shell on the silica nanoparticle core with a shell thickness of ~ 50nm. Copper in C-S CuSiNP is chelated by the silica silanol (Si–OH) groups in the silica matrix forming a weak Cu-Si complex. However, the reduction mechanism of copper is not well understood as no specific reducing agent was added externally during the synthesis process. However, it is possible that ethanol (which is produced after the TEOS hydrolysis) and silica (with –OH and –O⁻ groups) might have played a role of mild reducing agents. Atomic Absorption Spectroscopy (AAS) quantified the amount of copper to be 0.098 ppm in comparison to copper standards. **Figure 2a** shows the HRTEM micrograph of AgSiNG. The formation of silver nanoparticles ranging from 10-20 nm uniformly distributed in amorphous silica matrix (grey material in contrast) was confirmed by the HRTEM. The HRTEM – selected area electron diffraction pattern confirms crystallinity of the silver nanoparticles along with lattice planes of 2.36±0.05 Å for 111 and 2.04 ± 0.04 for 200 specific for Ag. Additionally we also identified 220 and 311 reflections in electron diffraction pattern that are specific for Ag crystals (**Figure 2b**). The formation of Ag nanoparticles can be accounted to addition of sodium hydroxide in the formulation that acted as a reducing agent[13] to reduce Ag^+ to Ag^0 leading addition of metallic Ag to produce Ag nanoparticles.

Antibacterial assays
    The growth inhibitory effects of C-S CuSiNP and AgSiNG against *E.coli* and *B.subtilis* were studied in liquid media (**Figure 3** and **Figure 4**). Bacterial growth with different concentrations of C-S CuSiNP and AgSiNG (0.49 to 9.8 μg mL⁻¹) was monitored after 24 hours of incubation at 37°C by measuring the optical density at 600nm using Teysche800 spectrophotometer. Since the turbidity of silica based material can interfere with the optical density reading, the background measurement was subtracted to calculate the final reading.
    C-S CuSiNP showed significant growth inhibition of two different strains of bacterium, gram-negative E.coli and gram-positive B.subtilis. Total inhibition was obtained at 9.8 ppm copper concentration for both the bacterium. C-S CuSiNP exhibited improved antibacterial efficacy in comparison to copper sulfate against E.coli as well as B.subtilis. This clearly shows that C-S CuSiNP has improved copper bioavailability when compared to copper sulfate with similar copper concentration. This could be attributed to the novel core-shell design, where majority of the copper is present in the shell. C-S CuSiNP is intermediate between "soluble" and "insoluble" copper compounds where the Cu ions are chelated in the silica matrix. This results in improved and sustained antibacterial activity in comparison to "soluble" copper sulfate.
    In case of AgSiNG higher growth inhibition could be seen in *E.coli* (**Figure 5**) than compared to *B.subtilis* (**Figure 6**), this could be attributed to the difference in the cell wall structure of the

bacteria. No statistically significant difference in the antibacterial efficacy was observed between the AgSiNG and the silver nitrate materials, suggesting that silica matrix served as a host matrix and did not compromise the antibacterial properties of silver. The growth inhibition can be seen at 2.4 ppm of silver in both AgSiNG and silver nitrate solutions. This corresponds to the value of MIC of silver[14]. Thus it can be seen that embedding silver in silica did not interrupt the antibacterial properties of silver ions.

Baclight live/dead cell staining was also done using ZEISS Axioskop2 confocal microscope to determine the cell viability of *E.coli* and *B.subtilis*. The images of live/dead cells with different concentrations of C-S CuSiNP were taken using a florescent microscope (**Figure 7**). Similarly images were taken for the bacteria incubated with different concentrations of AgSiNG using fluorescent microscope (**Figure 8**). The green filter (535nm) was used to view live cells and red filter (642 nm) was used to view dead cells. The amount of red cells was significantly greater than the green cells, confirming the bactericidal effect of the metallic based silica nanogel nanoparticle material.

CONCLUSIONS

Using a simple sol-gel method, core-shell copper loaded silica nanoparticle (C-S CuSiNP, ~450 nm) and silver loaded silica nanogel materials embedding 10-20 nm size crystalline silver nanoparticles have been successfully synthesized. In comparison to copper sulfate control, C-S CuSiNP material showed improved antibacterial properties against both *E. coli* (a gram-negative) and *B. subtilis* (a gram-positive) bacteria. This has been attributed to improved Cu bioavailability of C-S CuSiNP material where the core-shell design could have been played an important role. Antibacterial efficacy of silver did not compromise in AgSiNG material, suggesting that silica matrix served simply as a host material. The present study demonstrates that the silica matrix can be efficiently used as an inert delivery vehicle for metal based antibacterial active agents such as copper and silver as both nanoparticle and nanogel matrix formats.

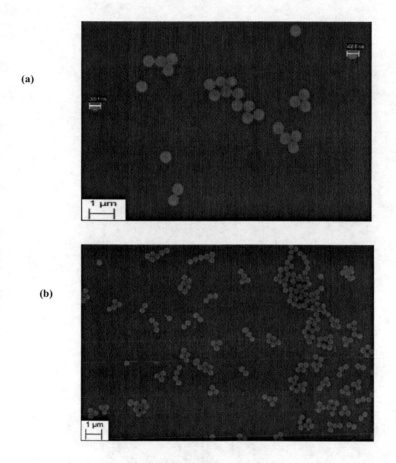

Figure 1: SEM images of SiNP ~380nm (a) and C-S CuSiNP ~450nm (b) showing particle size and morphology.

(a)

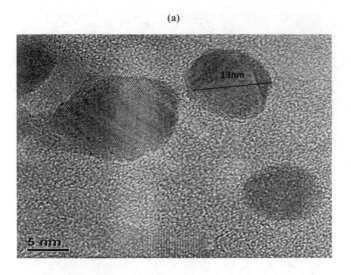

(b)

Figure 2: (a) HRTEM micrograph of AgSiNG showing the presence of silver nanoparticles in amorphous silica gel and (b) HRTEM – electron diffraction pattern of silver nanoparticles present in the AgSiNG material.

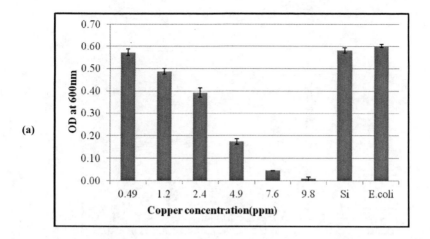

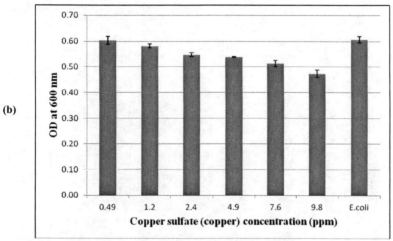

Figure 3: Histogram showing inhibition of *E.coli* in liquid media by C-S CuSiNP (a). Silica nanoparticle was used as negative control. Copper sulfate with equivalent metallic copper concentration was used as positive control as shown in (b).

(a)

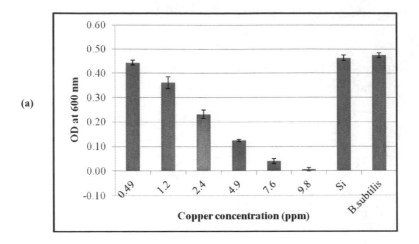

(b)

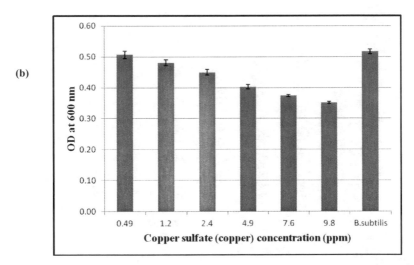

Figure 4: Histogram showing inhibition of *B.subtilis* in liquid media by C-S CuSiNP (a). Silica nanoparticle was used as negative control. Copper sulfate with equivalent metallic copper concentration was used as positive control as shown in (b).

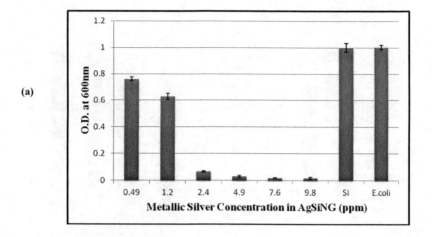

(a)

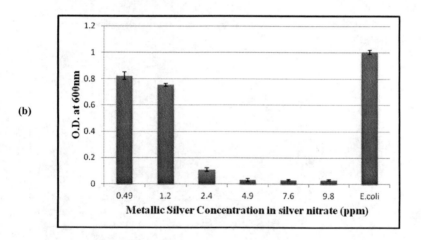

(b)

Figure 5: Histogram showing inhibition of *E.coli* in liquid media by AgSiNG (a). Silica nanogel was used as negative control. Silver nitrate with equivalent metallic silver concentration was used as positive control as shown in (b).

(a)

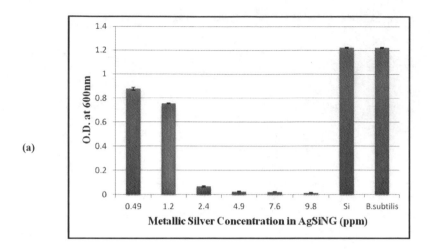

(b)

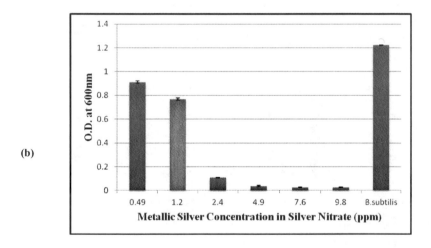

Figure 6: Histogram showing inhibition of *B.subtilis* in liquid media by AgSiNG (a). Silica nanogel was used as negative control. Silver nitrate with equivalent metallic copper concentration was used as positive control as shown in (b).

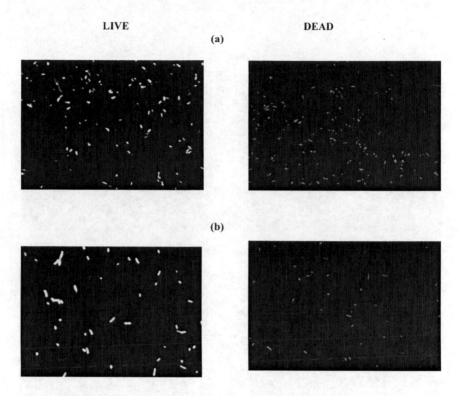

Figure 7: Fluorescent microscopy images of *E.coli* (a) and *B.subtilis* (b) showing live / dead cells on treatment with C-S CuSiNP material.

LIVE   DEAD

(a)

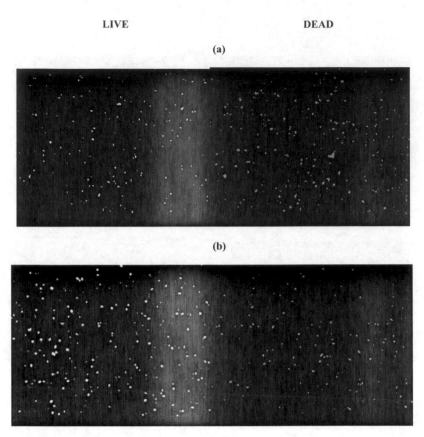

(b)

Figure 8: Fluorescent microscopy images of *E.coli* (a) and *B.subtilis* (b) showing live/dead cells on treatment with AgSiNG material.

ACKNOWLEDGEMENTS

The authors would like to acknowledge the Materials Characterization Facility (MCF) at the University of Central Florida (UCF) for the HRTEM and SEM characterization. Financial support from the Florida Department of Citrus (Grant # 186) and the Citrus Research and Development Foundation, Inc. (Grant # 328) is highly acknowledged.

REFERENCES
1. G. Borkow, J. Gabbay, Copper as a biocidal tool Current Medicinal Chemistry 12, 2163-2175 **(2005)**.
2. Y. Kobayashi, T. Sakuraba, Silica-coating of metallic copper nanoparticles in aqueous solution Colloids and Surfaces A: Physicochemical and Engineering Aspects 317, 756-759 **(2008)**.
3. P. Appendini, J. H. Hotchkiss, Review of antimicrobial food packaging Innovative Food Science & Emerging Technologies 3, 113-126 **(2002)**.
4. M. H. Freeman, C. R. McIntyre, A comprehensive review of copper-based wood preservatives: with a focus on new micronized or dispersed copper systems Forest Products Journal 58, 6-27 **(2008)**.
5. T. P. Schultz, D. D. Nicholas, A. F. Preston, A brief review of the past, present and future of wood preservation Pest Manage. Sci. 63, 784-788 **(2007)**.
6. A. Singh, V. Krishna, A. Angerhofer, B. Do, G. MacDonald, B. Moudgil, Copper Coated Silica Nanoparticles for Odor Removal Langmuir 26, 15837-15844 **(2010)**.
7. N. Voulvoulis, M. D. Scrimshaw, J. N. Lester, Alternative antifouling biocides Applied Organometallic Chemistry 13, 135-143 **(1999)**.
8. D. V. Quang, P. B. Sarawade, A. Hilonga, J.-K. Kim, Y. G. Chai, S. H. Kim, J.-Y. Ryu, H. T. Kim, Preparation of silver nanoparticle containing silica micro beads and investigation of their antibacterial activity Applied Surface Science 257, 6963-6970 **(2011)**.
9. P. Maniprasad, S. Santra, Novel copper (Cu) loaded core-shell silica nanoparticles with improved Cu bioavailability: synthesis, characterization and study of antibacterial properties Journal of Biomedical Nanotechnology (accepted on December 24, 2011)**(2012)**.
10. L. M. Rossi, L. F. Shi, F. H. Quina, Z. Rosenzweig, Stober synthesis of monodispersed luminescent silica nanoparticles for bioanalytical assays Langmuir 21, 4277-4280 **(2005)**.
11. M. Kawashita, S. Tsuneyama, F. Miyaji, T. Kokubo, H. Kozuka, K. Yamamoto, Antibacterial silver-containing silica glass prepared by sol,Äigel method Biomaterials 21, 393-398 **(2000)**.
12. S. K. Rastogi, V. J. Rutledge, C. Gibson, D. A. Newcombe, J. R. Branen, A. L. Branen, Ag colloids and Ag clusters over EDAPTMS-coated silica nanoparticles: synthesis, characterization, and antibacterial activity against Escherichia coli Nanomedicine: Nanotechnology, Biology and Medicine 7, 305-314 **(2011)**.
13. D. T. Sawyer, J. L. Roberts, Hydroxide ion: an effective one-electron reducing agent? Accounts of Chemical Research 21, 469-476 **(1988)**.
14. S. Egger, Lehmann, R. P., Height, M. J., Loessner, M. J., & Schuppler, M., Antimicrobial properties of a novel silver–silica nanocomposite material. Applied and Environmental Microbiology 75, 2973-2976 **(2009)**.

HOT WIRE AND SPARK PYROLYSIS AS SIMPLE NEW ROUTES TO SILICON NANOPARTICLE SYNTHESIS

M. R. Scriba[1], D.T. Britton[2] and M. Härting[2]

[1] National Centre for Nano-Structured Materials, CSIR, P O Box 395, Pretoria 0001, South Africa.
[2] Dept. of Physics, University of Cape Town, Rondebosch 7701, South Africa.

ABSTRACT

Monocrystalline silicon nanoparticles with a mean diameter of between 30 and 40 nm have been synthesised by hot wire thermal catalytic and spark pyrolysis at a pressure of 40 and 80 mbar respectively. For the production a mixture of the precursor gases, silane and diborane or silane and phosphine were used. While hot wire pyrolysis always results in multifaceted particles, those produced by spark pyrolysis are spherical. Electrical resistance measurements of compressed powders showed that boron doped silicon powders have a much higher conductivity than those doped with phosphorus. TEM and XPS analysis reveals that the difference in electrical resistivity between boron an phosphorus doped particles can be attributed to phosphorus dopants being located at the surface of the particles where an oxide layer is also observed. In contrast, boron doped particles are far less oxidised and the dopant atoms can be found in the core of the particle. The results demonstrate that hot wire and spark pyrolysis offer a new simple route to the production of monocrystalline doped silicon nanoparticles suitable for printed electrical devices.

INTRODUCTION

Silicon can be considered the semiconductor with the most useful properties for the manufacture of electronic devices such as individual transistors, integrated circuits and solar cells. However, due to the nature of the devices and their manufacturing processes, applications are generally limited to rigid mounting without the option of flexibility. Recently, printed electronics, based on silicon nanoparticle inks[1], has emerged as a new route to electrical device production. The technology currently allows the printing of silicon based devices on flexible media such as paper and polymers[2]. In these applications the printed nanoparticle network must be able to act as an extrinsic semiconductor, thus requiring the silicon nanoparticles to be doped and free of an oxide layer. Current bottom-up production processes for silicon nanoparticles include chemical vapour synthesis by laser or plasma pyrolysis[3] and laser ablation[4] at low pressure (high vacuum). In an attempt to find new simple routes for the synthesis of silicon nanoparticles , hot wire thermal catalytic pyrolysis (HWTCP)[5, 6] and spark pyrolysis (SP) have been applied. Whereas HWTCP is based on the well known technique of hot wire chemical vapour deposition, spark pyrolysis has been utilized for the first time in the production of silicon nanoparticles. In this process the spark can be defined as a sudden breakdown or ionisation of the precursor gas or vapour, due to a high electric field between two electrodes. Free electrons and to a lesser extent ions, form the current of charge carriers in the plasma of the spark[7] resulting in a near instantaneous heat source[8] reaching temperatures between 5000 °C and 22000 °C[9] in a few micro seconds. This high temperature heat source rapidly dissociates the silane and dopant gases during the production of doped silicon nanoparticles. Although electrically excited, spark pyrolysis has the most similarities with pulsed laser pyrolysis of precursor gases, which also has short heating and rapid cooling cycles[10].

EXPERIMENTAL

A basic vacuum system, in which the pressure, precursor gas flow rate and either the filament temperature or spark gap can be changed, was constructed to accommodate both HWTCP and SP configurations. Two sets of nanopowders were produced by each method, at a precursor flow rate of 50 sccm, firstly with a mixture of diborane (B_2H_6) and silane (SiH_4) and secondly with a mixture of phosphine (PH_3) and silane. Dopant gases were administered at ratios of 0.01, 0.1 and 1%. The production pressure for SP was 80 mbar and that of HWTCP 40 mbar. For all HWTCP synthesis experiments the filament temperature was maintained at 1800°C. For SP synthesis the spark was generated by continuously charging and discharging the high voltage power supply capacitor in a free running mode yielding an approximate energy of 0.6 J at a frequency of about 10 Hz. This frequency is determined by the breakdown voltage and power supply characteristics. Each production run of up to 20 minutes resulted in a layer of ochre to brown powder covering all internal surfaces of the reaction chamber. The powder was harvested after opening to air and stored in glass bottles without any special treatment.

The bulk crystallinity of the powders was investigated by x-ray powder diffraction (XRD), performed with a Panalytical PW3040/60 E'Pert Pro diffractometer, operated with CuKα radiation with a wavelength of 0.15406 nm. A JEOL JSM7500F analytical Field Emission scanning electron microscope (SEM) and a Thermo Scientific UltraDry silicon drift detector were used to collect energy dispersive x-ray spectroscopy (EDX) spectra at 5 keV, and the internal structure of the silicon nanoparticles was investigated using a JEOL 2100 TEM at 200kV acceleration voltage. A Physical Electronics Quantum 2000 x-ray photoelectron spectroscopy (XPS) system with monochromatic Al Kα radiation was used to investigate the surface composition of the nanoparticles. Lastly, the electrical characteristics of nanopowders were determined using a system specifically constructed for that purpose. For each sample, about 20 mg of powder was loaded into a plastic cylinder of 8.85 mm internal diameter and compressed between two stainless steel rods which also serve as the electrical contacts. The applied force was set with a calibrated load cell to 2.55 MPa and the change in separation of the pistons was measured by a micrometer. Current—voltage (I-V) curves were measured in current sweeping mode using a Keithley 4200 semiconductor characterisation system at room temperature (23°C). A similar electrical characterisation approach has been used by other groups on lead dioxide powders [11] and metal powders [12]. The conductivity of the compacted powders was determined from the measured resistance by

$$\sigma = \frac{h}{RA}, \qquad (1)$$

where h is the distance between the contacts, R the resistance, and A the cross sectional area of the compressed powder.

RESULTS

From the narrow, well defined peaks in the X-ray powder diffraction patterns of the silicon nanoparticles shown in figure 1, which do not show any contribution from an amorphous phase, the crystalline nature of all powders, is evident. The sharp peaks at 28.5°, 47° and 56° correspond to the (111), (220) and (311) crystal planes of crystalline silicon, and the three higher order peaks at 69° (400), 76.5° (331) and 88° (422) are also clearly visible.

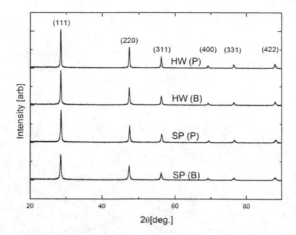

Figure 1. XRD patterns of silicon nanopowders produced with 1% dopant gas concentration by HWTCP (marked HW) and SP with diborane (marked B) and phosphine (marked P).

TEM studies clearly show that the nanoparticles are all monocrystalline irrespective of the production method or dopant gas. However, the TEM images reveal that HWTCP particles are always multifaceted (Fig 2a) and those produced by SP are always spherical (Fig 2b).

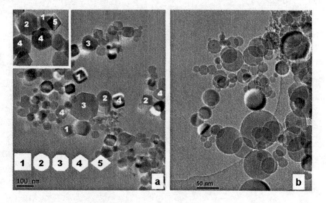

Figure 2. Example of silicon nanoparticles produced by (a) HWTCP (multifaceted) and (b) TCP (spherical). The numbered insets represent different projected views of the same basic morphology with a shape ranging from an octahedron to a truncated octahedron[13].

The mean diameter of all particles irrespective of the production method, pressure or dopant gas concentration is below 50 nm, however a small fraction of nanoparticles with diameters up to 150 nm is also present in these powders. At a higher TEM magnification a thin outer layer of a general amorphous nature on particles produced with phosphine (figure 3a) is revealed. From the TEM studies it was found that the thickness of this outer layer on the nanoparticles may be up to the equivalent of 3 of the (111) crystal planes, or about 0.6 nm, in the silicon nanoparticles produced at the highest dopant levels. Particles produced with diborane generally have lattice planes that extend to the surface (figure 3b) indicating the absence of disorder (amorphous shell).

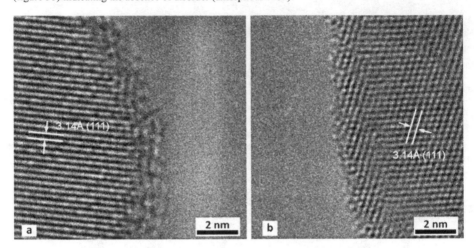

Figure 3. Surface structure of nanoparticles produced by SP with (a) 0.01% phosphine, and (b) 0.1% diborane. In both images the (111) crystal orientation is indicated.

Furthermore, by studying TEM images of all particles, it was observed that the amorphous surface region generally increases in thickness with an increase in dopant gas concentration for both phosphorus and boron doped silicon nanoparticles. The general composition of the powders was determined by EDX, and as shown by the inset example in figure 4, the particles comprise mainly silicon and oxygen with a small percentage of the dopant.

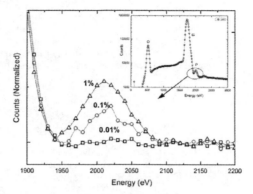

Figure 4. EDX spectra of particles produced by SP at 80 mbar showing an increase of phosphorus in the powder for an increase in phosphine concentration (from 0.01% to 0.1% and 1%) during production.

The increase in the intensity of the phosphorus peak (at about 2020 eV) shown in figure 4 indicates an increase in phosphorus atoms present in the particles with an increase in phosphine concentration during production (from 0.01 to 1%). The surface region of the nanoparticles was further investigated by XPS. Of special interest is the determination of the silicon oxide. For this reason, the silicon 2p peaks in the XPS spectrum at around 100 eV were analysed, after subtracting the background of the spectra, using the iterative Shirley method[14]. This region of the spectrum contains information on all silicon oxidation states. The contribution of the silicon sub-oxides was now determined by achieving a best least squares fit (Figure 5) from fitting for all 5 peaks corresponding to the different oxidation states of silicon[15].

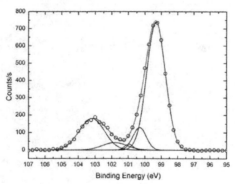

Figure 5. An example XPS spectrum of silicon nanoparticles showing the best least squares fit achieved, with 5 Gaussian peaks at about: 99.0 eV (Si), 100.0 eV (Si^{1+}), 100.65 eV (Si^{2+}), 101.5 eV (Si^{3+}), 103.0 eV (Si^{4+}).

To estimate the shell thickness on the silicon nanoparticles attributable to SiO_2, the ratio of the integrated intensities of the XPS peaks corresponding to Si and SiO_2 was used. For Si 2p photoelectrons the perpendicular escape depth in silicon is 2.11 nm. Using a maximum escape depth of 3.80 nm[16], an approximate depth of 1.9 nm in SiO_2 and 1.055 nm in silicon is probed by the XPS measurement. These values were used to calculate the average thickness of the SiO_2 layer for all particles, using the equation:

$$y = \frac{P_{SiO_2} \times R_X}{(R_X + 1)} \qquad (2)$$

where y is the thickness of the SiO_2 layer, P_{SiO2} is the photoelectron escape depth in SiO_2 and R_X is the ratio of the integrated intensities of the SiO_2 and silicon photoemission peaks. Using this method the estimated SiO_2 layer thicknesses for the silicon powders were determined and are presented in figure 6.

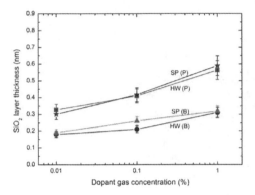

Figure 6. The estimated SiO_2 layer thickness of all particles at different dopant gas concentrations.

The results confirm the TEM observations: for particles produced with high phosphorus doping levels the fully oxidised silicon layer is up to 0.6 nm thick. Furthermore the SiO_2 content in the surface region of all nanoparticles produced in the presence of both dopant gases increases with increasing dopant concentration. The SiO_2 content in phosphorus doped particles is twice that of particles produced with boron as dopant. To determine the concentration of dopant atoms in the surface region of the doped silicon nanoparticles the boron 1s photoemission peak at about 188 eV[17] as well as the phosphorus P-O peak at 135 eV [18-20] and the combined P 2p peak, at 131 eV [20-23] were analysed. The ratio of the intensities of the boron peak and as the P-O bond, to the SiO_2 of all powders are shown in figure 7(a) and 7(b) respectively.

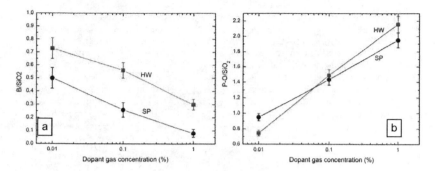

Figure 7. Comparison of dopant concentrations as a ratio of SiO_2 in all silicon nanopowders produced with (a) diborane (b) phosphine.

With an increase in dopant gas concentration, the number of dopant atoms in the outer 2-3 nm (max XPS probe depth) of nanoparticles decreases in the case of boron (figure 7a) and increases in the case of phosphorus (figure 7b). This observation is made for powders of both production methods.

Finally, the resistivity of all silicon nano-powders was measured using the method described earlier.

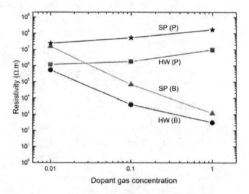

Figure 8. A comparison of the resistivity values of all powders produced by HWTCP and SP with varying concentrations of phosphine and diborane.

The resistivity curves shown in figure 8 reveal that powders produced by HWTCP and SP with diborane (marked as HW (B) and SP (B)) exhibit a decrease in resistivity with an increase in diborane concentration. In contrast, powders produced with phosphine (marked as HW (P) and SP (P)) show an increase in resistivity with an increase in phosphine concentration. Furthermore all powders produced by SP show a higher resistivity compared to powders produced by HWTCP, at the equivalent dopant concentrations.

DISCUSSION

Since the silicon nanoparticles produced in this study are ultimately intended for applications in printed electronic devices, the electrical characteristics of the powders are important. Furthermore, the resistivity will depend primarily on the composition of the particles, their doping concentration, the packing density (compression)[24], as well as the presence and thickness of an oxide layer on the surface of particles.

While particles produced with boron exhibit electrical activity those produced with phosphorus show an increase in resistivity with an increase in phosphine concentration. EDX analysis has indicated that this increase cannot be ascribed to the failure to incorporate phosphorous (fig 4) but rather to the increase in the thickness of a stoichiometric oxide layer on the surface of these particles (figure 6) with an increase in dopant gas concentrations. Furthermore, the phosphorous atoms are located near the particle surface. This is apparent from the substantial increase in P-O bonds (figure 7b) with the increase in the oxide layer thickness of these particles. In particles doped with boron, the maximum stoichiometric oxide detected at high dopant levels is below 0.3 nm (figure 6) and thus cannot be considered an oxide layer. Nevertheless, an increase in SiO_2 content with increased diborane gas levels is detected and corresponds to a decrease in the boron content in the surface region of these particles (figure 7a), indicating that the boron atoms are located inside the nanoparticles. The availability of charge carriers due to this successful doping is thus responsible for the decrease in resistivity with increasing doping levels.

The Cabrera-Mott mechanism [25] offers a possible way of interpreting the thicker oxide layer observed on particles produced with phosphine. A virgin silicon surface on a silicon nanoparticle will rapidly form a thin native oxide layer when exposed to air. Electrons may now tunnel through the thin oxide layer to the surface. Some oxygen atoms adsorbed on the surface will be ionized and acquire a negative charge, to form an electric potential V_{OL} across the oxide layer. Assisted by V_{OL}, oxygen ions diffuse through the silicon oxide layer to oxidise the silicon underneath [3]. From the model it can be deduced that in the case of phosphorous doped silicon nanoparticles there is a higher concentration of free electrons, induced by the phosphorus, especially if the phosphorus is primarily located at the surface. This results in an increase in V_{OL}, which boosts the transport of the oxygen ions through the silicon oxide layer, resulting in a more efficient oxidation. In boron doping, the low electron concentration reduces V_{OL}, and thus boron doped nanoparticles are less efficiently oxidised [3].

The difference in shape between silicon nanoparticles produced by HWTCP and SP is directly linked to their respective synthesis temperatures and cooling rates While the short, high temperature spark creates the conditions for rapid pyrolysis and rapid cooling of nanoparticles, the HWTCP process is characterised by gradual cooling. Both processes are graphically depicted in figure 9. Particles start to form in the supersaturated vapour, by homogeneous nucleation to form liquid silicon droplets [26, 27]. To maintain a minimum surface energy, the droplets are expected to be spherical at this stage. As the pressure in both processes is high, sufficient precursor species collide with the nucleating particles, resulting in accelerated growth. The difference in the shape of particles produced by SP and HWTCP originates from the difference in the rate of cooling, between the processes.

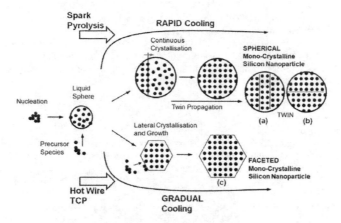

Figure 9   Silicon nanoparticle growth process for spherical and faceted particles.

The high temperature spark, which lasts for approximately 100 ns, creates the conditions for rapid pyrolysis and rapid cooling. Crystallisation of the spherical droplets start from a crystal nucleation site on its edge and propagates through the crystal as a crystallisation front [26] at the solid/liquid interface. The process results in spherical single crystal silicon nanoparticles. Similar results have been reported for laser pyrolysis [27, 28] and plasma synthesis with accelerated cooling [26, 29].

Continuing with figure 9, in contrast to spark pyrolysis, the HWTCP process has a continuous heat source, resulting in continuous pyrolysis and a moderate temperature gradient around the filament. The relatively slow growth rate allows the liquid silicon droplets to gradually cool as they move away from the filament. There is sufficient time and energy for atoms to find energy favourable growth sites. The crystal planes of the nanocrystals therefore start to grow laterally [30] to form faceted particles [31]. Furthermore, species in the vapour have enough energy to find an energy favourable sight and contribute to continued particle growth. Given ideal conditions an individual particle takes the shape with the lowest total surface energy, a truncated octahedron [13, 32-35]. However, the temperature gradient around the filament creates differences in cooling gradients, which is responsible for the different states of faceted shapes in particles produced by HWTCP as shown in figure 2(a).

CONCLUSION

Monocrystalline doped silicon nanoparticles have been produced by hot wire thermal catalytic pyrolysis and spark pyrolysis. The silicon nanoparticles are spherical when synthesised by SP and faceted for HWTCP. While the successful boron doping of the nanoparticles is evident from the decrease in resistivity with increasing dopant concentrations, the same does not apply to silicon nanoparticles doped with phosphorous. The resistivity of phosphorus doped nanoparticles increases with increasing phosphine concentration in the precursor gas. This effect is attributed to the formation of a silicon oxide on the surface of these particles up to a thickness of 0.6 nm. Furthermore, the phosphorus atoms are not incorporated in the core of the silicon nanoparticle but rather at the interface to the oxide layer where they are electrically inactive. The shape of the nanoparticles is governed by the synthesis temperature and cooling rate of the processes. The short-lived, high temperature and cyclic spark process creates the conditions for rapid pyrolysis and rapid cooling, resulting in spherical particles. In contrast the HWTCP process which has a gradual cooling profile results in the ordered growth of faceted nanoparticles.

ACKNOWLEDGMENT

The authors would like to acknowledge the project financial support of CSIR and the use of the Characterisation Facility at the DST/CSIR National Centre for Nano-Structured Materials, as well as the use of facilities at the University of Cape Town. XPS measurements were performed by W Jordaan at the National Metrology Institute of South Africa.

REFERENCES

[1]D.T. Britton and M. Härting, Printed Nanoparticulate Composites for Silicon Thick Film Electronics, *Pure and Applied Chemistry*, **78**(1723 (2006).
[2]M. Härting, J. Zhang, D.R. Gamota and D.T. Britton, Fully printed silicon field effect transistors, *Applied Physics Letters*, **94**(19), 193509-3 (2009).
[3]X.D. Pi, R. Gresback, R.W. Liptak, S.A. Campbell, and U. Kortshagen, Doping efficiency, dopant location, and oxidation of Si nanocrystals, *Applied Physics Letters*, **92**(12), 123102-3 (2008).
[4]F.E. Kruis, H. Fissan and A. Peled, Synthesis of nanoparticles in the gas phase for electronic, optical and magnetic applications--a review, *Journal of Aerosol Science*, **29**(5-6), 511-535 (1998).
[5]M.R. Scriba, C. Arendse, M. Härting and D.T. Britton, Hot-wire synthesis of Si nanoparticles, *Thin Solid Films*, **516**(5), 844-846 (2008).
[6]M.R. Scriba, D.T. Britton, C. Arendse, M.J. van Staden, and M. Härting, Composition and crystallinity of silicon nanoparticles synthesised by hot wire thermal catalytic pyrolysis at different pressures, *Thin Solid Films*, **517**(12), 3484-3487 (2009).
[7]N. Tabrizi, M. Ullmann, V. Vons, U. Lafont, and A. Schmidt-Ott, Generation of nanoparticles by spark discharge, *Journal of Nanoparticle Research*, (2008).
[8]Z.I. Ashurly, V.V. Gal and V.P. Malin, Temperature field in a pulsed discharge, *Journal of Engineering Physics and Thermophysics*, **20**(1), 45-49 (1971).
[9]R. Ono, M. Nifuku, S. Fujiwara, S. Horiguchi, and T. Oda, Minimum ignition energy of hydrogen-air mixture: Effects of humidity and spark duration, *Journal of Electrostatics*, **65**(2), 87-93 (2007).
[10]H. Hofmeister, P. Kodderitzsch and J. Dutta, Structure of nanometersized silicon particles prepared by various gas phase processes, *Journal of Non-Crystalline Solids*, **232-234**(182-187 (1998).
[11]H. Braun, K.J. Euler and P. Herger, Electronic conductivity of lead dioxide powder: separation of core and surface resistance of the particles, *Journal of Applied Electrochemistry*, **10**(4), 441-448 (1980).
[12]Y.P. Mamunya, H. Zois, L. Apekis and E.V. Lebedev, Influence of pressure on the electrical conductivity of metal powders used as fillers in polymer composites, *Powder Technology*, **140**(1-2), 49-55 (2004).
[13]S. Onaka, Geometrical analysis of near polyhedral shapes with round edges in small crystalline particles or precipitates, *Journal of Materials Science*, **43**(8), 2680-2685 (2008).
[14]J.E. Castle, H. Chapman-Kpodo, A. Proctor and A.M. Salvi, Curve-fitting in XPS using extrinsic and intrinsic background structure, *Journal of Electron Spectroscopy and Related Phenomena*, **106**(1), 65-80 (2000).

[15]M. Niwano, H. Katakura, Y. Takeda, Y. Takakuwa, N. Miyamoto, A. Hiraiwa, and K. Yagi, Photoemission study of the SiO[sub 2]/Si interface structure of thin oxide films on Si(100), (111), and (110) surfaces, *Journal of Vacuum Science and Technology A: Vacuum, Surfaces, and Films*, **9**(2), 195-200 (1991).

[16]K. Hirose, H. Nohira, K. Azuma and T. Hattori, Photoelectron spectroscopy studies of SiO2/Si interfaces, *Progress in Surface Science*, **82**(1), 3-54 (2007).

[17]X.J. Hao, E.C. Cho, C. Flynn, Y.S. Shen, S.C. Park, G. Conibeer, and M.A. Green, Synthesis and characterization of boron-doped Si quantum dots for all-Si quantum dot tandem solar cells, *Solar Energy Materials and Solar Cells*, **93**(2), 273-279 (2009).

[18]Y. Sano, W.B. Ying, Y. Kamiura and Y. Mizokawa, Annealing induced phosphorus protrusion into thin-oxide films from heavily phosphorus-doped silicon (100), *Thin Solid Films*, **516**(8), 1788-1795 (2008).

[19]W.B. Ying, Y. Mizokawa, K. Tanahashi, Y. Kamiura, M. Iida, K. Kawamoto, and W.Y. Yang, Evaluation of the initial oxidation of heavily phosphorus doped silicon surfaces using angle-dependent X-ray photoelectron spectroscopy, *Thin Solid Films*, **343-344**(393-396 (1999).

[20]Y. Mizokawa, W.B. Ying, Y.B. Yu, Y. Kamiura, M. Iida, and K. Kawamoto, Phosphorus redistribution in the surface region of heavily phosphorus doped silicon, *Applied Surface Science*, **100-101**(561-565 (1996).

[21]A.I. Kovalev, D.L. Wainstein, D.I. Tetelbaum, W. Hornig, and Y.N. Kucherehko, Investigation of the electronic structure of the phosphorus-doped Si and SiO$_2$:Si quantum dots by XPS and HREELS methods., *Surface and Interface Analysis*, **36**(8), 959-962 (2004).

[22]S.F. Yoon and R. Ji, Application of electron cyclotron resonance chemical vapour deposition in the preparation of hydrogenated SiC films: a comparison of phosphorus and boron doping, *Journal of Alloys and Compounds*, **261**(1-2), 281-288 (1997).

[23]X.J. Hao, E.C. Cho, G. Scardera, E. Bellet-Amalric, D. Bellet, Y.S. Shen, S. Huang, Y.D. Huang, G. Conibeer, and M.A. Green, Effects of phosphorus doping on structural and optical properties of silicon nanocrystals in a SiO2 matrix, *Thin Solid Films*, **517**(19), 5646-5652 (2009).

[24]A. Celzard, J.F. Maréché, F. Payot and G. Furdin, Electrical conductivity of carbonaceous powders, *Carbon*, **40**(15), 2801-2815 (2002).

[25]N. Cabrera and N.F. Mott, Theory of the Oxidisation of Metals., *Pep. Prog. Phys.*, **12**(163), (1948).

[26]C.R. Perrey and C.B. Carter, Insights into nanoparticle formation mechanisms, *Journal of Materials Science*, **41**(9), 2711-2722 (2006).

[27]F. Huisken, H. Hofmeister, B. Kohn, M.A. Laguna, and V. Paillard, Laser production and deposition of light-emitting silicon nanoparticles, *Applied Surface Science*, **154-155**(305-313 (2000).

[28]E. Borsella, M. Falconieri, S. Botti, S. Martelli, F. Bignoli, L. Costa, S. Grandi, L. Sangaletti, B. Allieri, and L. Depero, Optical and morphological characterization of Si nanocrystals/silica composites prepared by sol-gel processing, *Materials Science and Engineering B*, **79**(1), 55-62 (2001).

[29]B. Giesen, H. Wiggers, A. Kowalik and P. Roth*, Formation of Si-nanoparticles in a microwave reactor: Comparison between experiments and modelling, *Journal of Ianoparticle Research*, **7**(1), 29-41 (2005).

[30]A. Babat, C. Anderson, C.R. Perry, C.B. Carter, S.A. Campbell, and U. Kortshagen, Plasma synthesis of single-crystal silicon nanoparticles for novel electronic device applications, *Plasma Phys. Control. Fusion*, **46**(B97-B109 (2004).

[31]U.R. Kortshagen, L. Mangolini and A. Babat, Plasma synthesis of semiconductor nanocrystals for nanoelectronics and luminescence applications., *Journal of Ianoparticle Research*, **9**(39-52 (2007).

[32]A.S. Barnard and P. Zapol, A model for the phase stability of arbitrary nanoparticles as a function of size and shape, *The Journal of Chemical Physics*, **121**(9), 4276-4283 (2004).

[33]S. Onaka, A simple equation giving shapes between a circle and a regular N -sided polygon, *Philosophical Magazine Letters*, **85**(7), 359-365 (2005).

[34]K. Kimoto, Morphology and crystal structure of fine particles produced by a gas evaporation technique, *Thin Solid Films*, **32**(2), 363-365 (1976).

[35]Z.L. Wang and J.S. Yin, Self-assembly of shape-controlled nanocrystals and their in-situ thermodynamic properties, *Materials Science and Engineering A*, **286**(1), 39-47 (2000).

PLASMA-ASSISTED CHEMICAL VAPOR DEPOSITION OF Fe:TIO₂ FILMS FOR PHOTOELECTROCHEMICAL HYDROGEN PRODUCTION

Andreas Mettenbörger,[a] Vanessa Merod,[a] Aadesh P. Singh,[a] Helge Lemmetyinen[b] and Sanjay Mathur[a,*]

[a] Chair of Inorganic and Materials Chemistry, Department of Chemistry
University of Cologne, Greinstr. 6, Cologne D 50859, Germany

[b] Department of Chemistry and Bioengineering, Tampere University of Technology
P. O. Box 541, FI-33101 Tampere, Finland

*E-mail: sanjay.mathur@uni-koeln.de

ABSTRACT

Fe-doped TiO₂ (Fe:TiO₂) thin films were deposited onto fluorine-doped tin oxide (FTO, SnO_2:F), substrates via plasma enhanced chemical vapor deposition (PE-CVD) using a stoichiometric mixture of iron pentacarbonyl ($Fe(CO)_5$) and titanium isopropoxide ($Ti(OPr^i)_4$) as precursors. The evolution of surface morphology with changing process parameters was studied by scanning electron microscopy (SEM) and atomic force microscopy (AFM). UV-Visible spectral data showed a shift of the absorption edge towards the visible region and the band gap energy was found to gradually decrease with increasing dopant (Fe) concentration. Photoelectrochemical (PEC) measurements carried out on Fe:TiO₂ samples under 1 Sun illumination using 1M NaOH as electrolyte showed maximum photocurrent density and flatband potential of 1.26 mA/cm² and 0.68V/SCE, respectively at 5.0 at% Fe doping. Incident photon-to-electron conversion efficiency (IPCE) was found to increase from 0.75% to 4.0% upon increasing the dopant concentration upto 5.0 at%.

INTRODUCTION

Since the discovery of photocatalytic splitting of water in the early 1970's by Fujishima and Honda, TiO₂ remains a promising photoanode material for photoelectrochemical (PEC) application due to its optical and electrical properties, its stability against corrosion, industrial production and low cost.[1-3] However, the current application of TiO₂ for photoelectrochemical water splitting is limited due its large band gap (3.2 eV), which requires high energy photons (<388 nm) to generate electron–hole pair in the material.[4] However, this wavelength forms less than 4% of the solar spectrum and therefore, necessitates modification of TiO₂ to also harvest the visible part of solar radiation with the λ_{max} of ca. 600 nm.

In the past, several approaches, like metal ion doping,[5] anion doping,[6, 7] noble metal loading,[8] addition of electron donors,[9] metal ion-implantation[10] and self-doping that produces Ti^{3+} species[11] were examined to achieve a reduction of the band gap energy of TiO₂. Since the substitution of iron(III) ions into the TiO₂ lattice is favoured due to the comparable sizes of Fe^{3+} (78.5 pm) and Ti^{4+} (74.5 pm), several chemical preparation methods, such as sol-gel[12] or hydrothermal[13, 14], and gas phase approaches[15, 16] have been reported to obtain Fe-doped TiO₂. In this work, we report a single-step synthesis of Fe-doped TiO₂ via plasma enhanced chemical vapour deposition (PE-CVD), which is a powerful technique for the low-temperature deposition and modification of crystalline metal oxides at technically relevant-scale onto temperature sensitive materials like glass and polymers[17, 18]. For this purpose, a mixture of titanium isopropoxide and iron pentacarbonyl was used and the PE-CVD films were analysed towards their photoelectrochemical potential.

EXPERIMENTAL

In a typical experiment a mixture of 3 ml titanium isopropoxide (Across Organics) and calculated volume of iron pentacarbonyl (Across Organics) (0-5 at%) was prepared in a 10 ml round bottom glass flask and was connected to the PE-CVD reaction chamber (Plasma Electronic, Neuenburg, Germany). The precursor delivery system was maintained at 55°C to provide a sufficient vapour pressure. Conductive glass substrates (FTO, SnO_2:F), cleaned by sonification in water and isopropanol, were placed in the chamber, which was then pumped down to a pressure of ca. 0.5 Pa, before the precursors and 30 sccm of O_2 were introduced into the chamber, leading to a working pressure of ca. 3.0 Pa. After 1 h deposition with a plasma power of 70 W, poorly crystalline films were obtained which were post-annealed for 4 h at 500°C in air.

The FTO-samples were used to prepare photoelectrodes. A copper wire with a flat spiral at one end was dipped into silver conductive past, placed onto the uncoated part of the FTO and dried for one hour at room temperature. Afterwards the wire was fixed and covered with a 2-component epoxide glue. All edges were covered so that only the central part of the layer remained exposed. The thickness of the layers was measured on a silicon wafer with a Detac³ by Veeco profilometer. Scanning electron microscope (SEM) images and EDX data were collected with a field-emission SEM (Nova Nano SEM 430 (FEI)). A Perkin Elmer Lambda 950 UV-VIS- spectrometer was used for the optical measurements. The band edges of the samples were determined by performing a linear regression with extrapolation of the absorption band edge data. In order to get information about the topography and roughness a conductive AFM (I-AFM) XE-100 by Park Systems in the contact mode was used.

All photoelectrochemical measurements were conducted in a three-electrode electrochemical cell to obtain current-voltage (I-V) characteristics of undoped and iron doped TiO_2 films employed as working electrode under darkness and under one sun illumination having output intensity of 180 mW/cm² from a 150W xenon lamp (Newport, Model: 67005). 1 M NaOH electrolyte (pH = 13.6) was used as the electrolyte to record I-V characteristics at varying applied potential from +1.0 V/SCE (anodic bias) to -1.0 V/SCE (cathodic bias) with a scan rate of 10 mV/s for all samples.

Current-voltage curves needed to determine the voltage for the photo action spectra measurement were obtained by a SMU (source/monitor unit) E5272A with two Medium Power SMU modules E5282A by Agilent. The same was used for the photo-action spectra measurement. Light source was a Xe-lamp 75 W model 71208 by Newport and a Solar Simulator SSR by Luzchem. The sample was placed in a Faraday cage with illumination window made by TTY/Kemia. Finland.

RESULTS AND DISCUSSION

As-deposited films, TiO_2 and Fe:TiO_2, showed a colour change depending on the amount of iron doping from colourless (undoped) to reddish brown (5 at% in precursor mixture). The average film thickness measured by surface profilometry was found to be 70 nm (TiO_2) and 120 nm (Fe:TiO_2). The increase in film thickness is possibly due to the higher vapour pressure of $Fe(CO)_5$ when compared to $Ti(OPr^i)_4$, which can increase the overall precursor concentration in the gas phase leading to a higher growth rate.

The surface topography of undoped and Fe-doped TiO_2 films (Figure 1) showed a homogeneous surface consisting of spherical grains, whereas 5.0 at% Fe-doped TiO_2 sample displayed a smooth surface with few random agglomerates. These observations were confirmed by AFM analyses as evident in the 3D view of AFM images (Figure 1 c, d) of undoped and 5 at% Fe-doped TiO_2 thin films showed a decrease in the average surface roughness from 7.1 nm (undoped) to 0.5 nm (5 at% iron doping) with increasing doping concentration. The elemental film composition analysed by EDX confirmed the presence of iron (Table1). The Fe-content found in the Fe:TiO_2 films do not correspond to that of precursor mixture possibly due to differential vapor pressure and fragmentation pattern of $Ti(OPr^i)_4$[19] and $Fe(CO)_5$[20], are according to the equation 1 and 2 respectively.

Nevertheless, the trend suggests that more Fe is incorporated in the films with increasing Fe(CO)$_5$ concentration.

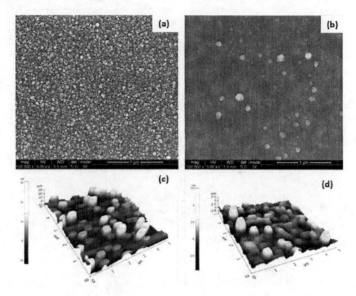

Figure 1. SEM (a and b) and AFM (c and d) images of undoped and 5.0 at% Fe doped TiO$_2$ thin films, respectivily.

$$Ti(OCH(HCH_3)_2)_4 \longrightarrow TiO_2 + 4C_3H_6 + 2H_2O \tag{1}$$

$$Fe(CO)_5 \longrightarrow Fe + 5CO \tag{2}$$

Table 1: Atomic percentage of elements found Fe:TiO$_2$ films by Energy Dispersive X-Ray Analysis (EDX).

Element	Undoped[at%]	2.0% Fe[at%]	3.5% Fe[at%]	5.0% Fe[at%]
O	10.56	10.98	14.76	9.81
Si	80.35	73.25	77.14	76.85
Ti	1.17	0.80	0.86	0.16
Fe	-	0.54	0.88	0.98

UV-Visible spectroscopy on FTO substrates showed a shift in absorption edge towards the visible region of solar spectrum with increasing doping concentration (Figure 3), indicating an improved solar energy harvesting capacity. The shift is caused by the additional energy level of the d-electrons of iron, which reduce the band gap [14]. With increasing doping concentration, the colour of the samples from pale to dark yellow (up to 4.0 at%) to brownish for the highest doping concentration

(5.0 at% Fe). The band gap energy calculated from the UV-Visible spectra were found to be in good agreement with values reported for other TiO_2 nanostructures[12]. In Fe:TiO_2 samples, it was observed that upon increasing the doping concentration the band edge shifted towards the visible region of the light due to additional energy levels provided by iron, which lie below the conduction band of TiO_2.

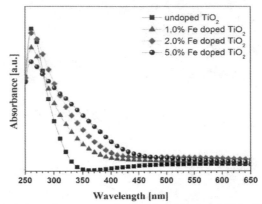

Figure 2. Optical absorption spectra of undoped and Fe-doped TiO_2 films with different doping concentrations.

Both undoped and Fe-doped TiO_2 thin films were used as photoelectrodes in PEC cell and current-voltage (I-V) characteristics were recorded under dark as well as under illuminated conditions. The observed current-potential characteristics reveal the expected pattern of a cell having n-type semiconducting characteristics. Figure 3 shows the photocurrent density versus applied potential curves obtained by subtracting the dark current from the light current for the undoped and Fe-doped TiO_2 thin films in 1M NaOH electrolyte solution. Fe incorporation in titanium dioxide was found to increase the photoelectrochemical response as observed in improved photocurrent density (Table 2). Undoped TiO_2 sample showed a photocurrent density of 0.10 mA/cm^2, whereas Fe:TiO_2 samples showed an improvement in photocurrent density. The 5.0 at% Fe-doping in TiO_2 exhibited the maximum photocurrent density ~1.26 mA/cm^2 at 0.6 V/SCE, which was 13 times higher as compared to that of undoped TiO_2. The variation in the photocurrent density with various doping level is mainly due to the shift in the band gap of TiO_2 towards the visible region and by changing its electronic properties through formation of shallow traps within the titania matrix.

Table 2. Photophysical properties of iron doped TiO_2 films at different Fe concentrations.

Doping concentration	Band edge position [nm]	Photcurrent density [mA/cm^2 at 0.6V/SCE]	Flatband Potential [mV/SCE]	depletion layer width [Å]	IPCE [% at 400 nm]
undoped	350	0.10	-0.54	147.1	0.75
1.0 at% Fe	375	0.15	-0.58	15.37	2.25
2.0 at% Fe	400	0.70	-0.62	4.07	3.65
5.0 at% Fe	450	1.00	-0.68	3.14	4.00

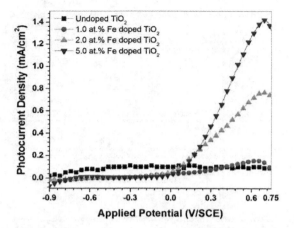

Figure 3. Photocurrent density versus applied potential for TiO₂ thin films doped at different doping concentration.

Figure 4 shows the Mott-Schottky plots ($1/C^2$ vs. applied potential) for undoped and Fe-doped TiO₂ using equation 3.

$$\frac{1}{C^2} = \left(\frac{2}{\varepsilon\varepsilon_o q N_D}\right)\left(V_{app} - V_{FB} - \frac{kT}{q}\right) \tag{3}$$

where ε_o is the permittivity of the vacuum, ε is the dielectric constant of the semiconductor and kT/q is the temperature dependent term. The flatband potential was found to increase from -0.54 mV/SCE (undoped TiO₂) to a value of -0.68 mV/SCE (5.0 at% Fe) with an increase of the doping concentration.

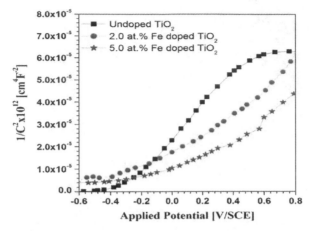

Figure 4. Mott-Schottky plots of undoped and Fe-doped TiO$_2$ films with different doping concentrations collected at 1 kHz frequency in 1M NaOH solution.

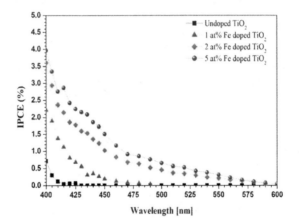

Figure 5. Incident photon to electron conversion efficiency (IPCE) for doped and undoped samples.

The depletion layer width (ω_d) was calculated (Table 2) with the simplified equation 4 suggested that depletion layer width drastically decreases with increasing iron concentration (147.1 Å for undoped TiO$_2$ and 3.14 Å for 5 at% iron doped films). A reduction of the depletion layer is favorable to suppress the recombination rate since the charges can be separated faster.

$$\omega_d = \frac{\varepsilon_0 \varepsilon}{c} \qquad (4)$$

Figure 5 shows the incident photon-to-electron conversion efficiency (IPCE) as a function of the wavelength (400 nm) as calculated from the photocurrent action spectra (Table 2). The monochromatic IPCE was found to increase with an increase of the Fe-doping concentration. The maximum IPCE value was obtained for highly (5.0 at%)-doped sample that displayed both enhanced photocurrent density and flatband potential. This observation is corroborated by the maximum lowering of band gap energy in 5 at% Fe-doped TiO$_2$ samples.

CONCLUSION

Fe:TiO$_2$ thin films deposited via PE-CVD technique were successfully utilized as photoelectrode material for photoelectrochemical water splitting applications. Structural, optical, electrical and photoelectrical properties revealed that with increasing dopant concentration, the absorption edge showed a red shift towards the visible region of solar spectrum. Photocurrent density and IPCE were found to increase with increasing dopant concentration, which makes this material a promising candidate for photoelectrochemical water splitting applications.

ACKNOWLEDGEMENTS

We gratefully acknowledge the University of Cologne and Nanommune Project (EC-FP7-NANOMMUNE- Grant Agreement No. 214281) for the financial support.

REFERENCES

[1] A. Fujishima, K. Honda, "Electrochemical Photolysis of Water at a Semiconductor Electrode", *Nature*, 238, 37-38, (1972).

[2] A. Fujishima, T.N. Rao, D.A. Tryk, "Titanium dioxide photocatalysis", *J. Photochem. Photobiol", C*, 1, 1-21, (2000).

[3] A.B. Murphy, P.R.F. Barnes, L.K. Randeniya, I.C. Plumb, I.E. Grey, M.D. Horne, J.A. Glasscock, "Efficiency of solar water splitting using semiconductor electrodes", *Int. J. Hydrogen Energy*, 31, 1999-2017, (2006).

[4] H. Tang, K. Prasad, R. Sanjines, P.E. Schmid, F. Levy, "Electrical and optical properties of TiO$_2$ anatase thin films", *J. Appl. Phys.*, 75, 2042-2047, (1994).

[5] H. Zhu, J. Tao, X. Dong, "Preparation and Photoelectrochemical Activity of Cr-Doped TiO2 Nanorods with Nanocavities", *J. Phys. Chem. C*, 114, 2873-2879, (2010).

[6] R. Asahi, T. Morikawa, T. Ohwaki, K. Aoki, Y. Taga, "Visible-Light Photocatalysis in Nitrogen-Doped Titanium Oxides", *Science*, 293, 269-271, (2001).

[7] S.U.M. Khan, M. Al-Shahry, W.B. Ingler, "Efficient Photochemical Water Splitting by a Chemically Modified n-TiO$_2$", *Science*, 297, 2243-2245, (2002).

[8] P.V. Kamat, M. Flumiani, A. Dawson, "Metal-metal and metal-semiconductor composite nanoclusters", *Colloids Surf., A*, 202, 269-279, (2002).

[9] A.A. Nada, M.H. Barakat, H.A. Hamed, N.R. Mohamed, T.N. Veziroglu, "Studies on the photocatalytic hydrogen production using suspended modified photocatalysts", *Int. J. Hydrogen Energy*, 30, 687-691, (2005).

[10] M. Takeuchi, H. Yamashita, M. Matsuoka, M. Anpo, T. Hirao, N. Itoh, N. Iwamoto, "Photocatalytic decomposition of NO under visible light irradiation on the Cr-ion-implanted TiO$_2$; thin film photocatalyst", *Catal. Lett.*, 67, 135-137, (2000).

[11] F. Zuo, L. Wang, T. Wu, Z. Zhang, D. Borchardt, P. Feng, "Self-Doped Ti^{3+} Enhanced Photocatalyst for Hydrogen Production under Visible Light", *J. Am. Chem. Soc.*, 132, 11856-11857, (2010).

[12]A.P. Singh, S. Kumari, R. Shrivastav, S. Dass, V.R. Satsangi, "Iron doped nanostructured TiO$_2$ for photoelectrochemical generation of hydrogen", *Int. J. Hydrogen Energy*, 33, 5363-5368, (2008).

[13]J. Ma, Y. Wei, W.-X. Liu, W.-B. Cao, „Preparation of nanocrystalline Fe-doped TiO$_2$ powders as a visible-light-responsive photocatalyst", *Res. Chem. intermed*, 35, 329-336, (2009).

[14]J. Zhu, W. Zheng, B. He, J. Zhang, M. Anpo, "Characterization of Fe–TiO$_2$ photocatalysts synthesized by hydrothermal method and their photocatalytic reactivity for photodegradation of XRG dye diluted in water", *J. Mol. Catal. A: Chem.*, 216, 35-43, (2004).

[15]G.R. Dai, *Sens. Actuators, B*, "A study of the sensing properties of thin film sensor to trimethylamine", 53, 8-12, (1998).

[16]X. Zhang, M. Zhou, L. Lei, "Co-deposition of photocatalytic Fe doped TiO$_2$ coatings by MOCVD", *Catal. Commun.*, 7, 427-431, (2006).

[17]S. Mathur, T. Ruegamer, "Transparent and Scratch-Resistant C:ZrO$_{(x)}$ Coatings on Polymer and Glass by Plasma-Enhanced Chemical Vapor Deposition", *Int. J. Appl. Ceram. Technol.*, 8, 1050-1058, (2011).

[18]J. Pan, R. Ganesan, H. Shen, S. Mathur, "Plasma-Modified SnO$_2$ Nanowires for Enhanced Gas Sensing", *J. Phys. Chem. C*, 114, 8245-8250, (2010).

[19]A. Rahtu, M. Ritala, "Reaction Mechanism Studies on Titanium Isopropoxide–Water Atomic Layer Deposition Process", *Chem. Vap. Deposition*, 8, 21-28, (2002).

[20]R. Tannenbaum, C.L. Flenniken, E.P. Goldberg, "Magnetic metal-polymer composites: Thermal and oxidative decomposition of Fe(CO)$_5$ and CO$_2$ (CO)$_8$ in a poly (vinylidene fluoride) matrix", *J. Polym. Sci., Part B: Polym. Phys.*, 28, 2421-2433, (1990).

CNT BASED NANOCOMPOSITE STRAIN SENSOR FOR STRUCTURAL HEALTH
MONITORING

A K Singh
Department of Aerospace Engineering,
Defence Institute of Advanced Technology (Deemed University),
Girinagar, Pune 411025, India.
Tel.: 020 24304173; Fax: 020 24389039
E-mail address: draksingh@hotmail.com; aksingh@diat.ac.in

ABSTRACT
    In the present endeavour an indigenously developed automated electric arc discharge method
was used to synthesize carbon nanotubes (CNT) by evaporating carbon electrodes in-water both with
and without argon atmosphere at the arcing electrodes. The synthesized CNT samples were
characterized at different stages of purification. TGA, Raman and FTIR spectra of CNT samples were
carried out to confirm the formation of CNTs. Formation of CNTs was further confirmed by SEM, and
HRTEM which also provided its morphology, growth pattern and size. For structural characterization
XRD patterns of synthesized CNTs were obtained. The CNTs obtained were about 10 nm in diameter
and about 0.5 μm in length. After purification and functionalization CNTs were mixed with
polymethyl methacrylate (PMMA) polymer to make CNT nanocomposite. The prepared CNT (5 %
loading)/PMMA nanocomposite was characterized for its electrical and chemical bonding properties.
The I-V characteristics of nanocomposite were linear between 0 V and 5V for temperature range 40 ^{0}C
to 70 ^{0}C. The peppered CNT/PMMA nanocomposite films were used for strain sensing application for
structural health monitoring by fixing it onto Aluminum (Al) and Stainless Steel (SS) cantilever test
beams. The strain sensor results closely match with standard strain gauge sensor under quasi static, as
well as, dynamic conditions.

INTRODUCTION
    Since the discovery of CNTs there has been wide interest in the synthesis, characterization and
exploration of wide variety of applications based CNTs and its composites. The reason is being that the
CNTs possess unusual and extraordinary mechanical, electrical, and thermal properties and have
shown potential to replace several existing technologies. Intrinsic coupling of electrical properties and
mechanical deformation in carbon CNTs and its composites makes them ideal candidates for future
multi-functional material systems that combine adaptive and sensory capabilities[1]. The mechanical and
electrochemical properties are coupled on CNTs, which is a characteristic of smart materials[2]. The
CNT based nonocomposites has been reported for wide range of applications i.e. new high strength
composites[3], EMI shielding[4], thermal management[5], smart materials[6], and detectors and actuators[7-9].
    The concept of creating both structural and functional multi-phase nanocomposites with
improved performance is currently under development in a wide variety of metallic, ceramic, and
polymeric matrices, although the emphasis to date has been on polymeric systems. Aerospace industry
focuses its research in producing multi-functional materials, driving design parameters being the
weight reduction with increased mechanical properties, as well as, monitoring their structural health by
means of sensing capability[10-18]. The sensing capability will primarily enhance safety and secondarily
reduce the intervals of maintenance in such materials. The electric resistance change has been used for
structural health monitoring by identifying internal damage of structures in composite materials. This

enabled full monitoring the structural health and establishing correlations between internal damage and change in resistance.

In view of above and for variety of other applications, great deal of interest has been generated in building highly sensitive strain sensors with CNT nanocomposites[10-19]. Initially, CNTs thin-film (called buckypaper) was reported for strain sensing. Strain induced change of electrical resistance (piezoresistivity) has been observed in these films. However, poor load transfer capability due to slippage among nanotubes limits its application as a strain sensor[18]. Nanocomposites provide a method to improve load transfer capability by using CNTs as filler in a polymer matrix. Additional advantage of composite based strain sensing is that it does not require expensive equipments for measurement. Additionally, use of polymer matrix improves the interfacial bonding between the CNTs which improves the strain transfer repeatability and linearity of the sensor. Highly strain-sensitive sensors can be obtained when the CNTs loading approaches the percolation threshold. At the percolation threshold, CNT clusters form a connected three-dimensional network resulting in a jump in the electrical conductivity. The percolation threshold in polymer-carbon composites is a complex phenomenon. To date, there has been no apparent consensus on the percolation threshold values for CNTs. For example, the values reported in the literature for typical CNT–epoxy nanocomposites vary from 0.002 to over 10 wt% [11-12]. Percolation threshold depends on the type of CNTs and the processing techniques used to produce the nanocomposites. It depends on the physico- chemical properties of both the polymer and filler, such as particle size, porosity, surface area as well as the composite processing conditions, such as temperature, solvent type [20]. Optimum performance of the sensors has been reported in the range of 3–10% w/w CNT loading[11,16]. Polymer matrix has an important role, high molecular weight polymers have long chains that are highly intermingled or spaghetti like in morphology as compared to the low molecular weight polymer that have shorter chains. More CNTs are needed to coat the high molecular weight polymers as well as form network structures through the polymer matrix. Several CNTs based nanocomposites strain sensors have been reported with a high gauge factor compared to conventional metallic foil strain gauges. CNTs/PMMA nanocomposites strain sensors have been reported with gauge factors up to 25[19].

It has been known fact that the properties of CNTs varies with the method of synthesis used. There are large numbers of techniques available to synthesize different types of carbon based nanostructures in the form of nanotubes, nanofibers etc., broadly classified as physical, chemical, hybrid and solvothermal/hydrothermal techniques[21-27]. The technique used depends upon the application of interest, type of nanostructure CNT/CNF material, quantity (scale of production), cost etc. A majority of high quality CNT's have been prepared using either the electric arc discharge or chemical vapor deposition.

Direct current(dc) arc plasma process for nanosynthesis has recently gained rapid momentum[26]. Simplicity of the setup, significantly high deposition rate, low production cost, high purity, property control coupled with chemical flexibility are the driving forces pushing this technique towards realization of pilot scale systems. Size and shape of the synthesized particles are of major fundamental and applied interest, as they determine the micro (electronic and quantum), as well as, the bulk behavior. Few of the crucial process parameters that govern the properties of nanotubes are the profiles of the chemical medium, fluid dynamics, applied field and thermal properties of the reacting gases, i.e., gas type, composition, flow rate, gas pressure, and temperature gradient in the plasma reactor[28-30]. The reactor geometry and the boundary wall temperatures also influence the formation of nanostructures. A brief account of various CNT synthesis methods is given by Huczko[31].

In principle, a single tube of CNT can form a sensor and provide important intrinsic sensing properties. However, until the technology to form regular array of single wires is fully developed, CNT based nanocomposite based sensor is an inappropriate device for mass production. The method of pasting CNT nanocomposite is a direct and easy way, but requires sequential processing of synthesis and collection of CNTs, as well as, dispersion in solutions or polymer, and printing on substrates.

Paper presents a systematic study from synthesis of CNTs, its characterization to device development. Arc discharge in water has been used for the synthesis of CNTs. The prepared CNT's were characterized at various stages of synthesis followed by the development of CNT-based composites and its use as strain sensor for structural health monitoring.

EXPERIMENTAL

Synthesis of CNT

Setup for Arc discharge in water (Sample VL1)

Laboratory scale arc discharge apparatus has been indigenously developed, for producing arc in-water for CNT production both with and without injecting argon gas at the arcing electrode. The anode was a graphite rod of 6.3 mm in diameter and cathode was a graphite rod of 12.7 mm in diameter (Alfa Aesar) and each 50 mm length. Water provides an oxygen free atmosphere for the reaction without need for an evacuated reactor. A specially fabricated 5mm thick Pyrex glass cylindrical container, placed vertical, filled with 2 L of de-ionized (DI) water of resistivity 18.2MΩ was used, having 40mm hole in its bottom to fix the anode holder. The upper end was open for removal of water and CNTs, as well as, for easy replacement of electrodes. It consists of a wooden cover which houses cathode holder and has a groove/counter-bore for upper open end of cylindrical container to fit in. A number of holes were drilled through it all along its surface. Brass cathode holder was fixed in central hole, all other holes were left open for hot air and vapors to escape through them. Container was filled with DI water till ¾ of the total chamber volume to ensure that graphite rod holder of cathode, as well as, anode was dipped in water. Arcing was initiated by moving electrodes by revolving wheel fitted outside on the electrode holder. The current was kept at 40A and arcing was done in bursts of 2 min each and bubbles were observed in water due to arcing. Once the full cathode rod was consumed, water was transferred to jars. This was allowed to cool and settle down for 2 hrs. It was followed by 30 min treatments of centrifugation to separate DI water & CNTs and finally drying in an oven. This method of synthesis was simple, as synthesis takes place at low temperature without vacuum conditions. Also overheating of cathode and anode holder was much less due to water medium.

Setup for arc discharge in water with injection of argon gas (Sample VL2)

The water medium provides necessary cooling but can cause more amorphous carbon being produced if it contains higher amount of dissolved oxygen. To overcome this drawback same setup was utilized with modification in cathode, as well as, in anode[32]. By injecting Ar gas through a hole in cathode, it was possible to isolate the arcing zone from water. The graphite anode was of 6 mm (4 mm tip) in diameter, and the cathode was of 13 mm in diameter with a hole of 11 mm diameter and 15mm in depth. The arc discharge was generated between these electrodes in the cathode hole. There is narrow hole (3 mm diameter) in the cathode to introduce Ar gas flow (2 litre/min) to exclude water from the arc zone. In this set-up arc plasma was isolated from the water by a graphite wall of 1mm thickness. This gas flow also serves to convey evaporated carbon from the hot arc spot to cold water. In the arc discharges, only the anode was consumed. A thin deposit, of approximately 0.5 mm thickness, was formed on the inner surface of the cathode hole, and this deposit was continuously peeled away into water by the arc pressure. During the arc discharge, fine powders floating on the water surface and bulky deposits including the cathode deposit settling down on the reactor bottom were obtained.

Preparation of nano-composite film for strain Sensor

For preparation of CNT composite PMMA have been used. PMMA require only simple laboratory instrumentation for the production of composite, and it is easier to uniformly combine CNTs with it. There are three main steps involved in the preparation of composite:

Dispersing CNT in a solvent:

Because of the small size of CNTs and their high surface area, fabricating functional macroscale materials has been a major challenge for applications such as sensing. Dispersion during composite processing was required to produce a suspension of independently separated nanotubes that can be manipulated into preferred orientations for large scale applications. Mechanical dispersion using ultrasonication or shear force mixing was used for dispersing the CNT's in dimethyl formamide (DMF) solvent, $2.5mg\ ml^{-1}$, and put in a sonicator bath for one hour, resulting fully dispersed CNT solution.

ii) Mixing PMMA and CNT

After the dispersion processing, the PMMA was added to a suspension of CNT's in DMF and mixed using shear force with hand and magnetic stirrer at 70 °C for 1 hr.

iii) Casting of the resulting CNT polymer composite in a mould

After dispersion and mixing process, the liquid was cast in a self fabricated Teflon mould to form a film and it was initially cured in a vacuum oven at room temperature for 30 min to remove air. It was then fully cured in a vacuum at 120 °C for 12 h to evaporate the solvent and to anneal the CNTs in the binding material. Then the composite film was peeled off from the Teflon mould. Same process was followed to prepare CNT/PMMA nanocomposites of different CNT loadings. The results presented in this paper are for 5% CNT loading in PMMA.

Characterization

The morphology of different samples was examined using JOEL, JSM-6360A and Philips XL-30 Analytical Scanning electron microscope (SEM). High resolution transmission electron microscope (HRTEM) of samples was performed using FEI, TECNAI F 30 HRTEM with 300Kv FEG to characterize morphology of synthesised material. The crystal structure of different CNT nanostructure samples were analyzed using Xpert PRO Panalytical Powder X-ray diffractometer (XRD) in the scanning range of $20\text{-}80^0$ (2θ) using Cu K_α radiations with wavelength 1.5045Å. The samples after filtration and grounding were investigated using a Du Pont 952 thermo-gravimetric apparatus with a thermo-balance, equipped with a computer control unit for the recording of TGA. Raman Spectroscopy was conducted on Laser Raman Spectrometer SPEX 1403 Spectrometer. FTIR spectra of CNT samples have been done on to confirm the formation of CNT's. Electrical characterization of CNT/PMMA composite film was done by fixing film on a glass substrate of approximate size 35mm x 15mm and by painting two 2-3mm narrow conducting silver strips at approximate gap of 5-6mm on top of the composite film. For the purpose a controlled temperature laboratory set up was used. The dimensions of this CNT/PMMA composite sensor were selected according to the dimensions of a commercial (metallic) strain gage. Both gages, commercial and based on CNTs, were bonded symmetrically on the center of the aluminum specimen, simulating a side by side configuration.

Experimental set-up for strain sensing

The prepared CNT based composite was tested for strain sensing by fixing it to an Al and SS test beams. An cantilever test beam was used as it was a simple structure for modelling and testing the response of the sensor. One end of the beam was rigidly clamped to a Iron cuboids block, as shown in Figure 1. Each strain sensor was tightly bonded using epoxy adhesive to ensure that the superglue

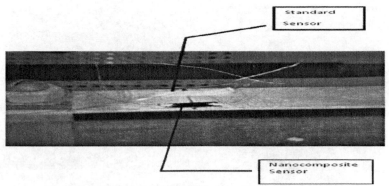

Figure 1. Experimental CNT /PMMA composite strain sensing setup.

makes a stiff bond to transfer the strain across the sensor without any slippage. The sensors were connected to cu lead wires with silver conducting paint to reduce the contact resistance. The change of resistance of each strain sensor was measured by Data Acquisition System (DT80 De Logger) with respect to the displacement due to bending of the cantilever beam. Full bridge circuit was used to measure change in resistance to minimise the errors due to temperature variations.

RESULT & DISCUSSION

SEM, HRTEM and EDAX of samples

Figure 2 a,b show the scanning electron micrographs (SEM) of synthesized CNT samples VL1 and VL2. SEM confirms the formation of CNT which are of about 10nm in diameter and 0.5μ mm in length. The synthesized CNT's were further analyzed using HRTEM, Figure 3, which confirms formation of good quality of MWNT i.e. the tubes have smooth wall and no observable defects. SEM of CNT/PMMA composite as shown in Figure 4, indicates good alignment of CNT and structural bonding with the polymer. From the TEM image, the MWNT are estimated to have diameter of ~10nm and ~0.5 μm in length.

The EDAX measurement of samples VL1 and VL2 are shown in Figure 5 which shows dominance of carbon component as compared to very less atomic percentage of platinum and other impurities of graphite electrode, as well as, oxygen content.

Figure 2. SEM of CNT samples VL1 (a) and VL2 (b).

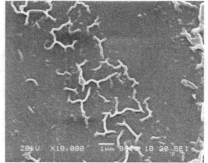

Figure 3. HRTEM of sample CNT sample VL1.   Figure 4. SEM of CNT/PMMA composite based on CNT sample VL1.

Thermal Gravimetric Analysis (TGA)

TGA analysis can be used for purity analysis of samples. Figures 6 show the TGA of CNT samples VL1 and VL2. These runs were carried out from 30 ^{0}C to 800 ^{0}C at a rate of 10 ^{0}C/min. Oxidation of both samples begin at about 550° C and with considerable weight loss of approximately, 97 wt% in the range from 550° C to 710° C as shown in figure. The carbonaceous carbon present in the sample is converted to CO/CO_2 which gets evolved in air. It is important to note that VL1 and VL2 has almost similar amount of metal and carbon impurity as can be seen from the Figure 6. The residual weights (which is usually contributed by metallic particles arises from electrodes) of the CNT samples are about 3 % which is in agreement with EDAX of samples.

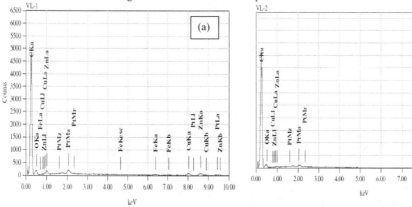

Figure 5. EDAX spectra of CNT samples VL1(a) and VL2 (b).

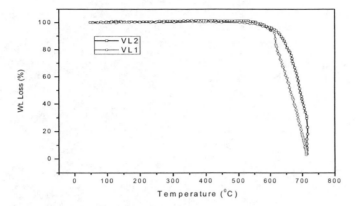

Figure 6. Purity analysis based on TGA of synthesized CNT samples VL1 and VL2.

FTIR of samples

Figure 7 a and b show the FTIR spectra recorded in transmittance mode for CNT samples VL1 and VL2. Peaks at 1113 cm^{-1}/1116 cm^{-1}, 1632/1602 cm^{-1}, 1742 cm^{-1}, 2924 cm^{-1} and 3440 cm^{-1} have been observed. The peak at about 3400 cm^{-1} is attributed to the presence of –OH groups and molecular water. The peak at 1742 confirms not only formation of CNT but also indicated its type i.e. MWNT. Strong bands are evident in the region 1600, in particular at 1602 cm^{-1} and 1621 cm^{-1} for MWNT. To verify that the PMMA is covalently bonded to CNT's, FTIR spectra of the nanocomposite sample was carried out as shown in Figure 8, the band at 1748 cm^{-1} is attributed to the ester bond (O–C¼¼O) in PMMA and the CH2 stretch appears at around 2900 cm^{-1}. Taken together, all this evidence confirms the formation of linkage between the free CNTs and PMMA.

XRD of samples

Representative XRD of CNT sample VL1 is shown in Figure 9. The diffraction peaks were observed at 26.3°, 43.7° and 53.8° which corresponds to CNT (101), CNT (102) and CNT (110) plane. Most prominent peak was observed at 26.3° indicating good crystalline quality of the prepared CNT.

Raman Spectroscopy

The most intense Raman features in CNT's are the G-peak at about 1600 cm^{-1} and the 2D-peak at about 2700 cm^{-1}[33]. The G-peak is due to first-order Raman scattering by the doubly degenerate zone center optical phonon mode and the 2D-peak is associated with second-order scattering by zone-boundary phonons. Figure 10 a shows the room temperature Raman spectra of CNT sample produced by arc discharge in water. The first peak observed at 1320 cm^{-1} (D band) while second peak at 1600 cm^{-1} (G band). The first overtone has been observed at 2650 cm^{-1}. The intensity of D is less than G band peak which implies formation of larger amount of crystalline carbon compound, like MWNT. Intensity ratio of G/ D peak is much higher which represent higher purity of CNTs and of larger length. The appearance of the D peak at 1320 cm^{-1} is attributed to the defects [34]. Figure 10b shows Raman spectra of VL2 sample. The first peak at 1350 cm^{-1} (D band) while second peak at 1600 cm^{-1} (G band) and a peak at 2650 cm^{-1} (first overtone) have been seen. The intensity of D is much less than G band peak which implies good crystalline carbon structure indicating formation of SWNT alongwith

MWNT. Also the splitting of G band implies existence of SWNT. The similar peaks for CNT analysis in Raman spectra has been reported by earlier researchers [34-36].

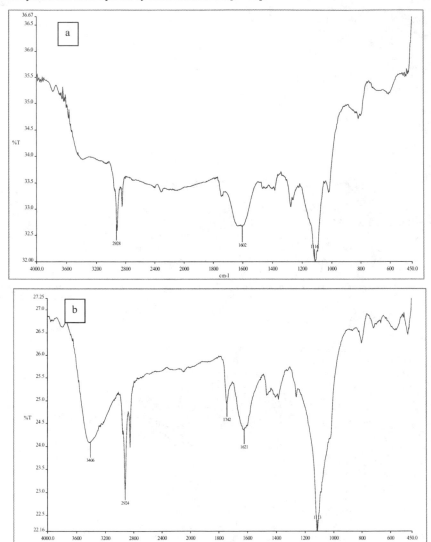

Figure 7. FTIR Spectra of CNT Samples (a) VL1 and (b) VL2.

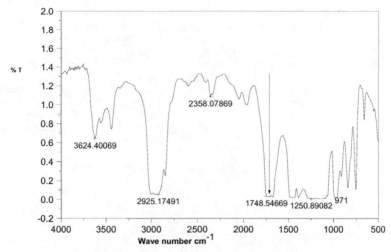

Figure 8. FTIR of CNT/ PMMA composite based on VL1 sample.

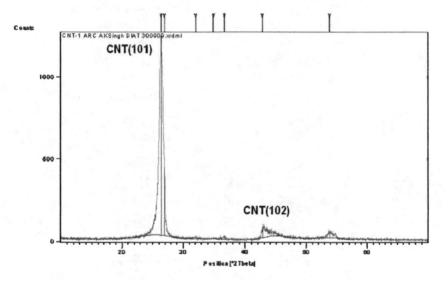

Figure 9. XRD of CNT sample VL1.

I-V characteristics of CNT/PMMA composite

Since both the samples VL1 and VL2 have similar characteristics only sample VL1 has been used for preparing CNT/PMMA composite. The change of resistance of the CNT/PMMA nanocomposite above room temperature from 40 ° C to 100 ^{0}C under applied voltage of 0 to 5 V is shown in Figure 11. The I-V characteristics of nanocomposite are linear between 0V and 5V for temperature close to 40 ^{0}C, and become nonlinear from 50 ^{0}C and above at 4V bias voltage. Nonlinear behavior became significant from 90 ^{0}C. The corresponding I-V behaviour indicates that the films have metallic (resistive) behaviour close to room temperature. At bias voltage of 4 V, for temperature above 70 ^{0}C, current started saturating showing non-ohmic behaviour. As reported CNTs can be either metallic or semiconducting, depending on the chirality of the tube, the nanocomposite film simultaneously has metallic (high conductivity) and semiconducting properties.

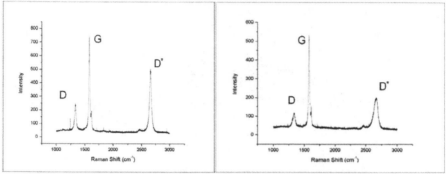

Figure 10. Raman spectra of CNT sample (a) VL1 and (b) VL2.

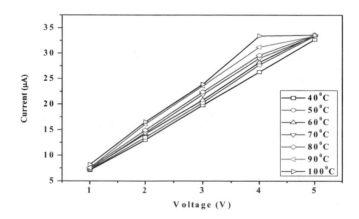

Figure 11. I-V characteristics of CNT/PMMA nanocomposite sample based on VL1.

Strain Sensor Response

Static Response

Figure 12 shows the change in resistance due to application of stress on Al test cantilever. Both the change in resistance of standard strain sensor and sample VL1 based nanocomposite sensors have similar waveform. In Figure 13 the change in resistance with initial resistance $(\Delta R/R)$ of nanocomposite strain sensor with respect to change of strain has been shown. The antisymmetric behavior, seen from the Figure 14, can be attributed to tunneling effect. Hu et al [17] has also reported similar anti-symmetric behavior behaviour. Anti-symmetric behavior corresponding to tensile and compressive strains was also reported in the sensors made from SWNTs [37].

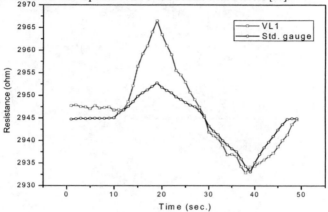

Figure 13. Static response of CNT/PMMA nanocomposite on application of stress.

Dynamic Response

In Figure 15 rise time of PMMA/CNT nanocomposite strain sensor and strain gauge has been shown. Dynamic behavior of nanocomposite strain sensor and strain gauge has been shown in Figure 16 under 1 and 2Hz i/p excitation applied from a mechanical shaker on SS cantilever test beam. Figure 17 shows the change in resistance of nanocomposite strain sensor and individual sensor with respect to strain, measured when the beam was deflected and allowed to oscillate freely. The sensor response in these Figures is almost identical to the output of the standard strain sensor, Type NK-7B having Gauge Factor 2.82 (M/s Rohits & Co.), which means that the composite sensor measures the strain signal from the structure in real time and without much distortion. It showed consistent piezoresistive behavior under repetitive tensile and compressive loading, and good resistance stability.

CONCLUSIONS

Indigenously developed arc in water was used to synthesize good crystalline CNTs. However, we did not observe significant difference in the crystalinity of two samples (Vl1 and Vl2). The electrical resistivities of the films were measured in situ using laboratory-designed fixtures and data acquisition system. The developed sensors exhibited a broad range of sensitivity, the upper limit showing nearly an order of magnitude increase compared to conventional, resistance-type strain gages. The Film sensor followed the free vibration pattern similar to pattern of standard reference strain sensor demonstrating good static and dynamic response. It showed consistent piezoresistive behavior under repetitive straining and unloading, and good resistance stability. The piezoresistance effect is promising for strain sensor. The sensor can be bonded to a surface, such as aircraft skin, to monitor the macroscopic strain in the structure. The high sensitivity sensing capability can be utilized to detect cracks in critical areas and therefore prevent catastrophic structural failure. Research is in progress on proof-of concept of the developed sensor as a low cost, reliable, high performance strain sensing device for structural health monitoring.

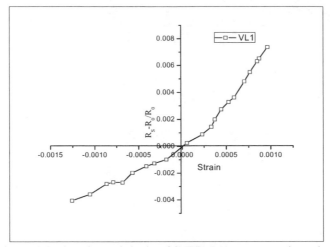

Figure 14. Strain resistance behavior of CNT/PMMA nanocomposite strain sensor.

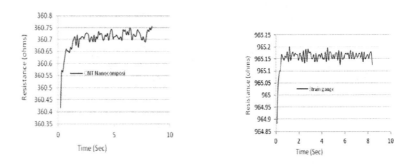

Figure 15 . Rise time of PMMA/CNT nanocomposite strain sensor and strain gauge.

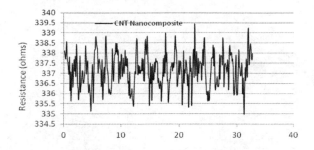

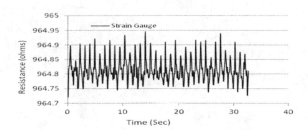

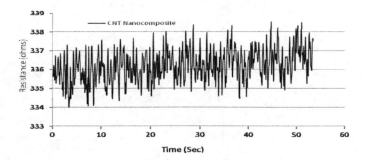

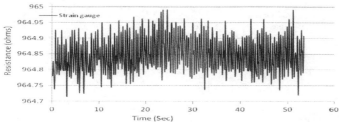

Figure 16. Dynamic response of CNT nanocomposites sensor and strain gauge under 1 and 2Hz i/p excitation from mechanical shaker applied to SS cantilever test beam.

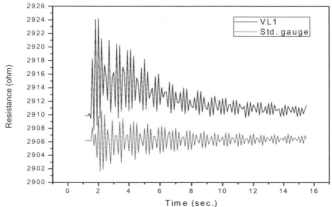

Figure 17. Comparison of dynamic strain response of free vibration for an CNT/PMMA composite strain gauge and standard strain gauge.

ACKNOWLEDGEMENTS

Authors are thankful to Vice Chancellor, DIAT, Girinagar, Pune for granting permission to publish this work. Thanks are also due to Physics Department, University of Pune, Pune; C-MET, Pune; NCL, Pune, HEMRL, Pune for characterization of samples.

REFERENCES

[1]N. Chakrabortya, B. Dharmasenab, Vivek T. Rathoda, Vinayak Naikc, D. Roy Mahapatraa, Implementation of a Wireless Mote with Nano-Composite Sensor for Structural Health Monitoring, *XVI lational Seminar on Aero space Structures (IASAS)* , November, 1-6, 2009.
[2]Inpil Kang, Yun Yeo Heung, Jay H. Kim, Jong Won Lee, Ramanand Gollapudi, Srinivas Subramaniam, Suhasini Narasimhadevara, Douglas Hurd, Goutham R. Kirikera, Vesselin Shanov, Mark J. Schulz,, Donglu Shi, Jim Boerio, Shankar Mall, and Marina Ruggles-Wren, Introduction to carbon nanotube and nanofiber smart materials, *Composites: Part B*, **37,** 382–394(2006).

[3]R. Andrews, D. Jacques, A.M. Rao, T. Rantell, F. Derbyshire, Y.Chen, J. Chen and R.C. Haddon, Nanotube Composite Carbon Fibers, *Applied Physics Letters*, **75**,1329–1331(1999).
[4]Zhuangjun Fan, Guohua Luo, Zengfu Zhang, Li Zhou, Fei Wei, Electromagnetic and microwave absorbing properties of multi-walled carbon nanotubes/polymer composites, *Materials Science and Engineering B*, **132**, 85–89(2006).
[5]Quoc Ngo, Brett A. Cruden, Alan M. Cassell, Megan D. Walker, Qi Ye, Jessica E. Koehne, M. Meyyappan, Jun Li2, and Cary Y., Thermal Conductivity of Carbon Nanotube Composite Films, *Mat. Res. Soc. Symp. Proc.* **812**, F3.18.1-6(2004).
[6]X. Li, C. Levy, A. Agarwal, A. Datye, L. Elaadil, A. K. Keshri and M. Li, Multifunctional Carbon Nanotube Film Composite for Structure Health Monitoring and Damping, *The Open Construction and Building Technology Journal, 3,* 146-152(2009).
[7]Basudev Pradhan, Ryan R. Kohlmeyer, Kristina Setyowati, Heather A. Owen and Jian Chen, advanced carbon nanotube/polymer composite infrared sensors, *Carbon* **47**, 1686 –1692(2009).
[8]Christele Bartholome, Alain Derre, Olivier Roubeau, Ceeile Zakri and Philippe Poulin, Electromechanical properties of nanotube-PVA composite actuator, bimorphs, *Ɉanotechnology* ,**19** 325501(6pp) (2003).
[9]Deuk Yong Lee, Il-Seok Park, Myung-Hyun Lee, Kwang J. Kimb and Seok Heo, Ionic polymer–metal composite bending actuator loaded with multi-walled carbon nanotubes, *Sensors and Actuators A,* **133**, 117–127(2007).
[10]K. J. Loh, J. P. Lynch, B. S. Shim and N. A. Kotov, Tailoring Piezoresistive Sensitivity of Multilayer Carbon Nanotube Composite Strain Sensors, *Journal of Intelligent Material Systems And Structures,* **19**, 747-764(2008).
11Gang Yin, Ning Hu, Yoshifumi Karube3, Yaolu Liu, Yuan Li and Hisao Fukunaga, A carbon nanotube/polymer strain sensor with linear and anti-symmetric piezoresistivity, *Journal of Composite Materials*, **45**(12) 1315–1323 (2011).
[12]G. T. Pham, Y. B. Park, Z. Liang, Z. Chuck and B. Wang, Processing and modeling of conductive thermoplastic/carbon nanotube films for strain sensing", *Composites Part B Engineering,* **39**(1), 209-216 (2008).
[13]M. Knite, V. Tupureina, A. Fuith, J. Zavickis and V. Teteris, Polyisoprene-multi-wall carbon nanotube composites for sensing strain, *Materials Science and Engineering C*, **27**, 1125-1128 (2007).
[14]Jin-Ho Kim, Jong min Kim, Inpil Kang, Inpil Kang, A Strain Positioning System Using Carbon Nanotube Flexible Sensors for Structural Health Monitoring, *2010 International Conference on Ɉanotechnology and Biosensors IPCBEE, Vol.2, IACSIT Press, Singapore, (2011).*
[15]H. Zhao, Y. Zhang, P. D. Bradford, Q. Zhou, Q. Jia, F. G. Yuan and Y. Zhu, Carbon nanotube yarn strain sensors, *Ɉanotechnology* **21**, 305502-06 (2010).
[16]J.R. Bautista-Quijano, F. Avilés, J.O. Aguilar and A. Tapia, Strain sensing capabilities of a piezoresistive MWCNT-polysulfone film, *Sensors and Actuators A: Physical*, **159**, 135-140(2010).
[17]N. Hu, Y. Karube, M. Arai, T. Watanabe, C. Yan, Y. Li, Y. Liu, H. Fukunaga, Investigation on sensitivity of a polymer/carbon nanotube composite strain sensor, *Carbon*, **48**, 680–687(2010).
[18]I. Kang, M. Schulz, J. Kim, V. Shanov and D. Shi, A Carbon nanotube Strain Sensors for Structural Health Monitoring, *Smart Materials and Structures*, **15**, 737-748 (2006).
[19]Yang Liu, Shantanu Chakrabartty, Dimitris Stamatis Gkinosatis, Amar K.Mohanty , and Nizar Lajnef, Multi-walled Carbon Nanotubes/Poly(L-lactide) Nanocomposite Strain Sensor for Biomechanical Implants, IEEE 1-4244-1525-X/119-122 (2007).
[20]By Jing Li, Peng Cheng Ma, Wing Sze Chow, Chi Kai To, Ben Zhong Tang, and Jang-Kyo Kim,Correlations between Percolation Threshold, Dispersion State and aspect ratio of carbon nanotubes, *Adv. Funct. Mater.*, **17**, 3207–3215(2007).

[21]J. Kong, A. M. Cassell and H. Dai, Chemical vapor deposition of methane for single-walled carbon Nanotubes, *Chemical Physics Letters,* **292,** 567–574(1998).

[22]D Pradhan, M Sharon, Carbon nanotubes, nanofilaments and nanobeads by thermal chemical vapor deposition process, *Mater Sci. Eng B,* **96,** 24–28(2002).

[23]S. N. Bondi, W. J. Lackey, R. W. Johnson, X. Wang and Z. L. Wang, Laser assisted chemical vapor deposition synthesis of carbon nanotubes and their characterization, *Carbon,* **44,** 1393–1403(2006).

[24]Y. Gogotsi, A. L. Joseph and M. Yoshimura, Hydrothermal Synthesis Of Carbon Nanotubes, *Proceedings Of Joint Sixth International Symposium On Hydrothermal Reactions (ISHR)* Japan, 350-355, 2001.

[25]W. Wang, , S. Kunwar and Z. F. Ren, Low Temperature Solvothermal Synthesis Of Multiwall Carbon Nanotubes, *Janotechnology,* **16,** 21–23 (2005).

[26]M. Caraman, G. Lazar and M. Stamate, Arc Discharge Instalaion For Fullerene Production, *Roman. Journal Physics,* **53,** (2008) 273-278.

[27]Karthikeyan, S. Mahalingam and M. Karthik, Large Scale Synthesis of Carbon Nanotubes, *Journal of Chemistry,* **6,** 1-12(2009).

[28] S. Ghorui, S. N. Sahasrabudhe, A. K. Tak, N. K. Joshi, N. V. Kulkarni, S. Karmakar, I. Banerjee, S. V. Bhoraskar, and A. K. Das, Role of Arc Plasma Instability on nanosynthesi, *IEEE Transactions on Plasma Science,* **34,** (2006) 121-127.

[29]M. V. Antisari , R. Marazzi, R. Krsmanovic, Synthesis of multiwall carbon nanotubes by electric arc discharge in liquid environments, *Carbon,* **41,** 2393–2401(2003).

[30] L. P. Biro, Z. E. Horvath, L. Szalmas, K. Kertesz, F. Weber, G. Juhasz, G. Radnoczi and J. Gyulai, Continuous carbon nanotube production in underwater AC electric arc, *Chemical Physics Letters,* **372,** 399–402(2003).

[31] A. Huczko, Synthesis of aligned carbon nanotubes, *Applied Physics A,* **74,** 617–638(2002).

[32] N. Sano, Low-cost synthesis of single-walled carbon nanohorns using the arc in water method with gas injection, *J. Phyics. D: Applied Physics,* **37,** L17–L20(2004).

[33]A. K. Pal, R. K. Roy, S. K. Mandal, S. Gupta and B. Deb, Electrodeposited carbon nanotube thin films, *Thin Solid Film,s,* **476,** 288–294(2005).

[34]G. D. Yuan, W. J. Zhang , Y. Yang, Y. B. Tang, Y. Q. Li, J. X. Wang, X. M. Meng, Z. B. He, C.M.L. Wu, I. Bello, C.S. Lee and S.T. Lee, Graphene sheets via microwave chemical vapor deposition, *Chemical Physics Letters,* **467,** 361–364(2009).

[35]C. Liu, H. Cheng, H. T. Cong, F. Li, G. Su, B. L. Zhou, and M. S. Dresselhaus, Synthesis of Macroscopically Long Ropes of Well-Aligned Single-Walled Carbon Nanotubes, *Advanced Materials,* **12,** 1190-1192(2000).

[36]S. S. Xiea, W. Z. Li, Z. W. Pan, B. H. Chang and L. F. Sun, Carbon nanotube arrays, *European Physics Journal D,* **9,** 85-89(1999).

[37]I. Kang, M. J. Schulz, J. W. Lee, G. R. Choi, J. Y. Jung and J. B. Choi, A carbon nanotube smart material for structural health monitoring. *Solid State Phenomenon,* **120,** 289–296(2007).

PHOTOCATALYTIC DEGRADATION OF WASTE LIQUID FROM BIOMASS GASIFICATION IN SUPERCRITICAL WATER WITH SIMULTANEOUS HYDROGEN PRODUCTION OVER CdS SENSITIZED $Na_2Ti_2O_4(OH)_2$

Wendong Tang, Dengwei Jing, Simao Guo, Jiarong Yin, Liejin Guo*
State Key Laboratory of Multiphase Flow in Power Engineering
Xi'an Jiaotong University, Xi'an,ShaanXi 710049, P. R. China

Abstract: Photocatalytic water splitting by solar light and biomass gasification in supercritical water both are considered as the most promising routes for renewable hydrogen production. The waste liquid from biomass gasification in supercritical water contains various organic compounds which can act as electron donors in photocatalytic system. In the present work, we report the hydrogen production by the anaerobic photocatalytic reforming of waste liquid from biomass gasification in supercritical water over CdS sensitized $Na_2Ti_2O_4(OH)_2$ nanotubes, Various reaction parameters were investigated, such as different gasification temperature and the pH value of the reaction solution. It was demonstrated that gasification in supercritical water and photocatalytic water splitting can be coupled together.

1. INTRODUCTION

The depletion of fossil fuels and the worldwide environmental problems make hydrogen an attractive alternative energy source. Hydrogen is a storable and environmentally friendly fuel. However, about 95% of hydrogen consumed in the world is produced from fossil fuels such as natural gas, petroleum and coal.[1] Thus to develop an alternative hydrogen production method is also important to solve the environment and energy problems.

Supercritical water gasification of biomass and photocatalytic water splitting under solar light both have been considered as the most promising routes for renewable hydrogen production.[2-9]

During the development of photocatalytic hydrogen production technology, numerous oxide[10,11] and sulfide[12,13] semiconductors have been developed since Fujishima and Honda reported the photoelectrocatalytic production of hydrogen from water on the TiO_2 semiconductor electrode.[14] However, the rapid recombination of photogenerated electron/hole pairs leads to the lower efficiency of photocatalysis.[15] Sacrificial agents, such as Na_2S, Na_2SO_3, alcohol, and organic acid were used to react irreversibly with the photogenerated holes, which prevented the recombination.[16-24]

On the other hand, supercritical water gasification is an effective way of wet biomass utilization. Biomass energy of low quality can be converted to hydrogen energy of high quality by supercritical water gasification.[3,4] Nevertheless waste liquid from supercritical water gasification still contains various kinds of organic compounds most of which are oxygen-containing ones, including ketones, hydroxybenzenes, esters, methoxyaromatics, and so on.[25] To couple these two systems supercritical water gasification of biomass and photocatalytic water splitting

* Corresponding author. Tel.: +86 29 82663895; fax: +86 29 82669033.
  E-mail address: lj-guo@mail.xjtu.edu.cn (L. Guo).

together, we tried photocatalytic hydrogen production in waste liquid from biomass gasification in supercritical water. Here organic compounds in the waste liquid were supposed to act as the sacrificial agents in the photocatalytic system.

2. EXPERIMENT

2.1 Preparation

According to the good performance of TiO_2 in organic sacrificial system,[16,22-24,26] $Na_2Ti_2O_4(OH)_2$ nanotube was chosen as p hotocatalyst. It was reported that $Na_2Ti_2O_4(OH)_2$ was existing in TiO_x state[27] with large specific surface area for strong physical and chemical adsorption and one-dimensional nanostructure for efficient separation and transportation of charges.[28] CdS was used as a sen sitizer in order to make it responsible to visible light.

The $Na_2Ti_2O_4(OH)_2$ nanotubes were prepared by t he facile hydrothermol method.[28,29] rutile TiO_2 was stirred in dense NaOH aqueous solution until a white suspension was obtained, then the suspension was aged at 150 °C for 48h. Then $Na_2Ti_2O_4(OH)_2$ nanotubes were obtained. The following synthesis step was a partial ion-exchange reaction. Cd^{2+} ions were introduced into the nanotube walls by sonicating $Na_2Ti_2O_4(OH)_2$ nanotube powder in ammonia containing $Cd(CH_3COO)_2$ solution. The obtained white $Cd^{2+}/Na_2Ti_2O_4(OH)_2$ slurry were washed with deionized water to remove the excess Cd^{2+}. Then the $Cd^{2+}/Na_2Ti_2O_4(OH)_2$ powder were transformed into yellow $CdS/Na_2Ti_2O_4(OH)_2$ by allowing a sufficient amount of Na_2S dropwise to react with the $Cd^{2+}/Na_2Ti_2O_4(OH)_2$ solution.

2.2 Characterization

X-ray diffraction patterns (XRD) of the synthesized samples were obtained from a PANalytical X'pert MPD Pro diffractometer using Ni-filtered Cu Kα irradiation. Transmission electron microscope (TEM) images were captured on a JEOL 2010 microscope using copper mounted lacey carbon grids. The chemical oxygen demand (COD) of the reaction solution was measured by PC MultiDirect with photometric method, and the results were calculated in $mg\cdot L^{-1}$.

2.3 Photocatalytic activity test

Photocatalytic water splitting was performed in a Pyrex photoreactor. 0.1 g of photocatalyst was dispersed in an aqueous solution (9 ml waste liquid in 90 mL aqueous solution). The waste liquid was obtained from biomass gasification in supercritical water under 25 MPa with a continuous flowing system.[2,3,30-32] A 500 W Xe lamp was used as the light source for hydrogen production. The amount of evolved H_2 was estimated using an on-line TCD gas chromatograph (TDX-01 column, N_2 as carrier gas). The blank experiments showed no appreciable hydrogen evolution in the absence of irradiation or photocatalyst.

3. RESULT AND DISCUSSION

3.1 Morphology and structure of CdS/Na$_2$Ti$_2$O$_4$(OH)$_2$

The morphology of the obtained products can be seen in Fig. 1. It is observed that CdS/Na$_2$Ti$_2$O$_4$(OH)$_2$ is successfully synthesized in the hydrothermal process. Titanium nanotubes are uniformly distributed with 150-200 nm in length and 8-10 nm in width. The hollow inner pore is around 3 nm with open tube ends, which is in accordance with the characteristics reported by the literature.[33] From Fig. 1b, CdS nanoparticles of 3-6 nm are well dispersed around the nanotubes.

Fig. 1 TEM micrograph of (a) Na$_2$Ti$_2$O$_4$(OH)$_2$ and (b) CdS/Na$_2$Ti$_2$O$_4$(OH)$_2$

Fig. 2 shows the XRD patterns of the composite photocatalyst. It can be seen that Na$_2$Ti$_2$O$_4$(OH)$_2$ exhibits an orthorhombic structure.[34,35] The observed peaks CdS/Na$_2$Ti$_2$O$_4$(OH)$_2$ sample show the presence of cubic crystalline phase CdS (JCPDS No. 10-0454)[36], decorated on Na$_2$Ti$_2$O$_4$(OH)$_2$. During the formation of nanotubes, the rutile phase of TiO$_2$ precursor still remains in the structure.

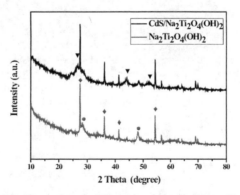

Fig. 2 XRD patterns of prepared samples. (●) nanotube orthorhombic structure; (♦) rutile TiO$_2$; (▼) cubic CdS.

3.2 Effect of waste liquid from different biomass gasification

Fig. 3 compares the photocatalytic performance of CdS/Na$_2$Ti$_2$O$_4$(OH)$_2$ in different dilute waste liquid solution from glucose cellulose and glycerol gasification in supercritical water respectively. As can be seen from this figure, hydrogen can be generated in the waste liquid over CdS/Na$_2$Ti$_2$O$_4$(OH)$_2$, which verified the possibility to couple photocatalytic hydrogen production and biomass supercritical gasification together.

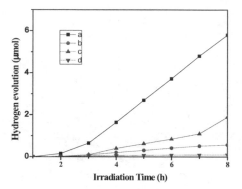

Fig. 3 Photocatalytic hydrogen generation in dilute waste liquid from different biomass gasification. (a) waste liquid from 15 wt.% glycerol gasification in 650 °C supercritical water; (b) waste liquid from 10 wt.% glucose gasification in 650 °C supercritical water; (c) waste liquid from 10 wt.% cellulose gasification in 650 °C supercritical water; (d) deionized water.

Compared with deionized water, waste liquid did enhance the photocatalytic activity of CdS/Na$_2$Ti$_2$O$_4$(OH)$_2$. As supposed, some of these organic compounds in waste liquid may react irreversibly with the photogenerated holes, which prevented the recombination.

Moreover, after supercritical water gasification, different biomass may generate waste liquid of different components. As shown in Figure 3, CdS/Na$_2$Ti$_2$O$_4$(OH)$_2$ performed better in the dilute waste liquid obtained from glycerol gasification. Glucose and cellulose molecules themselves are larger and more complicated than glycerol. Therefore after supercritical gasification, glucose and cellulose would generate much more kinds of organic compounds with more complex molecules in the waste liquid than glycerol. Some organic compounds themselves were difficult to degrade. And larger molecules were not conducive to adsorption for reaction. As a result, waste liquid from glycerol gasfication may be favorable to photocatalytic hydrogen production.

3.3 Effect of waste liquid from glycerol gasification in different supercritical condition

Hydrogen generation in waste liquid from glycerol gasification in different supercritical condition were investigated over $CdS/Na_2Ti_2O_4(OH)_2$.

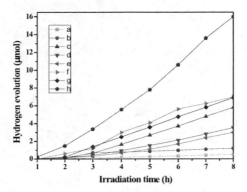

Fig. 4 Photocatalytic hydrogen generation in dilute waste liquid from glycerol gasification in different supercritical condition. (a) waste liquid from 4 wt.% glycerol gasification in 650 °C supercritical water; (b) waste liquid from 8 wt.% glycerol gasification in 650 °C supercritical water; (c) waste liquid from 10 wt.% glycerol gasification in 650 °C supercritical water; (d) waste liquid from 15 wt.% glycerol gasification in 650 °C supercritical water; (e)waste liquid from 20 wt.% glycerol gasification in 650 °C supercritical water; (f) waste liquid from 10 wt.% glycerol with 0.5 wt.% NaOH gasification in 600 °C supercritical water; (g) waste liquid from 10 wt.% glycerol with 0.5 wt.% Na_2CO_3 gasification in 600 °C supercritical water; (h) waste liquid from 10 wt.% glycerol with 0.5 K_2CO_3 gasification in 600 °C supercritical water;

As shown in Fig. 4, different supercritical condition affects the hydrogen generation rate. Compared with other concentration of glycerol origin aqueous solution, more hydrogen was obtained in waste liquid from 15 wt.% glycerol gasification in 650 °C supercritical water. However, if K_2CO_3, Na_2CO_3 or NaOH was added into glycerol origin aqueous solution, then after supercritical gasification, the obtained waste liquid was more suitable for photocatalytic reaction over $CdS/Na_2Ti_2O_4(OH)_2$. The highest hydrogen generation rate reached 1.39 $\mu mol \cdot h^{-1}$ in the waste liquid obtained from glycerol with K_2CO_3 gasification.

3.4 Effect of solution pH

The effect of solution pH on the hydrogen evolution from waste liquid was investigated, where the pH was adjusted by HCl and NaOH.

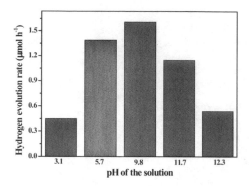

Fig. 5 Effect of solution pH on photocatalytic hydrogen generation in dilute waste liquid from glycerol
gasification in supercritical water

Fig.5 clearly shows that solution pH could strongly affect the hydrogen generation. Weak basic and weak acid conditions are favorable for photocatalytic hydrogen production, while low hydrogen evolution rate appears at pH 3.1 and pH 12.3. This result is in accordance with the literatures, which also report better photocatalytic performance in weak basic and weak acid environment.[37,38] The maximum hydrogen evolution rate 1.61 μmol•h^{-1} was obtained at pH 9.8.

In this complex organic solution, the effect of pH on the hydrogen evolution is very complicated, involving the changes of chemical state of various organic compounds and the surface charge. In the acid condition, organic compounds are mainly in molecular form which is not conducive to adsorption; in basic condition, if pH value was larger than the isoelectric point of catalyst, catalyst surface is negatively charged. Thus the electrostatic repulsion will inhibit adsorption of ionized organic groups. Therefore the maximum activity should be observed in the pH value near 7.

3.5 Degradation of waste liquid with simultaneous hydrogen production

Fig. 6 describes the change in COD value of the dilute waste liquid solution within 4 h irradiation. With the increase of reaction time, the COD value decreased. After 4 h irradiation, COD of the solution declined by 33.6%, demonstrating that hydrogen generation and degradation of waste liquid can be carried out simultaneously in this system. The blank experiments showed no appreciable COD change in the absence of photocatalyst under the same condition, which indicated that no selfdecomposition happened and the above process was a photocatalytic reaction.

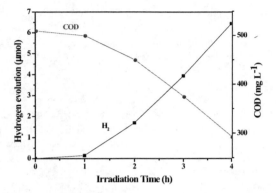

Fig. 6 Amount of photocatalytic H_2 generation and COD of reaction solution as a function of reaction time. Reaction condition: pH adjusted to 9.8, 90 mL of reaction solution containing 9 mL waste liquid from 10 wt.% glycerol with 0.5 wt.% K_2CO_3 gasification in 600 °C supercritical water.

According to the redox equilibrium, the decreased COD value should correspond to hydrogen evolution yield. From fig. 6, after 4 hours reaction, COD value reduced by 171 mg•L^{-1}, which means 1220 μmol e⁻ should have participated in the reducing reaction. However, only 6.5μmol H_2 was generated during this time. This can be ascribed to two reasons: First, some organic compounds are volatile. Thus they may easily spread into the gas phase, while COD value is only tested in the liquid solution. Second, some impurity gas like CO, CH_4 may be generated during the reaction which also consumes large amount of e⁻¹.[21] Thus prompt separation of organic gas is necessary to enhance the hydrogen evolution yield.

4. CONCLUSION
It was first time reported that hydrogen could be produced in the waste liquid from biomass gasification in supercritical water over CdS/$Na_2Ti_2O_4(OH)_2$ The photocatalytic performance was tested in dilute waste liquid from different supercritical condition. If K_2CO_3, Na_2CO_3 or NaOH was added into glycerol aqueous solution, then after supercritical gasification, the obtained waste liquid was more suitable for photocatalytic reaction. After adjusting the pH of the solution to 12, hydrogen evolution rate reached maximum 1.61 μmol•h^{-1} in dilute waste liquid from supercritical gasification glycerol and K_2CO_3 aqueous solution. The energy conversion efficiency is only 0.008%. Such low hydrogen generation rate and energy conversion efficiency could be ascribed to the complex components of waste liquid. It is crucial to identify these components and remove harmful ingredients to enhance the efficient.

ACKNOWLEDGEMENT
The authors gratefully acknowledge the financially supported of the National Natural Science Foundation of China (Contracted No. 50821064) and the National Basic Research Program of China (Contracted No. 2009CB220000).

REFERENCE
[1] Balat H, Kirtay E. Hydrogen from biomass - Present scenario and future prospects. International Journal of Hydrogen Energy 2010;35 (14): 7416-7426.
[2] Lu Y, Guo L, Ji C, et al. Hydrogen production by biomass gasification in supercritical water: A parametric study. International Journal of Hydrogen Energy 2006;31 (7): 822-831.
[3] Guo L, Lu Y, Zhang X, et al. Hydrogen production by biomass gasification in supercritical water: A systematic experimental and analytical study. Catalysis Today 2007;129 (3-4): 275-286.
[4] Kruse A. Supercritical water gasification. Biofuels Bioproducts & Biorefining-Biofpr 2008;2 (5): 415-437.
[5] Chen XB, Shen SH, Guo LJ, et al. Semiconductor-based Photocatalytic Hydrogen Generation. Chemical Reviews 2010;110 (11): 6503-6570.
[6] Kudo A, Miseki Y. Heterogeneous photocatalyst materials for water splitting. Chemical Society Reviews 2009;38 (1): 253-278.
[7] Jing D, Guo L, Zhao L, et al. Efficient solar hydrogen production by photocatalytic water splitting: From fundamental study to pilot demonstration. International Journal of Hydrogen Energy 2010;35 (13): 7087-7097.
[8] Khan SUM. Efficient Photochemical Water Splitting by a Chemically Modified n-TiO$_2$. Science 2002;297 (5590): 2243-2245.
[9] Murdoch M, Waterhouse GIN, Nadeem MA, et al. The effect of gold loading and particle size on photocatalytic hydrogen production from ethanol over Au/TiO2 nanoparticles. Nature Chemistry 2011;3 (6): 489-492.
[10] Kapoor MP, Inagaki S, Yoshida H. Novel zirconium-titanium phosphates mesoporous materials for hydrogen production by photoinduced water splitting. Journal of Physical Chemistry B 2005;109 (19): 9231-9238.
[11] Konta R, Ishii T, Kato H, et al. Photocatalytic activities of noble metal ion doped SrTiO$_3$ under visible light irradiation. Journal of Physical Chemistry B 2004;108 (26): 8992-8995.
[12] Tawkaew S, Fujishiro Y, Yin S, et al. Synthesis of cadmium sulfide pillared layered compounds and photocatalytic reduction of nitrate under visible light irradiation. Colloids and Surfaces a-Physicochemical and Engineering Aspects 2001;179 (2-3): 139-144.
[13] Zong X, Han J, Ma G, et al. Photocatalytic H2Evolution on CdS Loaded with WS$_2$ as Cocatalyst under Visible Light Irradiation. The Journal of Physical Chemistry C 2011; 115 (24): 12202–12208.
[14] Fujishima A, Honda K. Electrochemical photolysis of water at a semiconductor electrode. Nature 1972;238 (5358): 37-38.
[15] Ni M, Leung MKH, Leung DYC, et al. A review and recent developments in

photocatalytic water-splitting using TiO_2 for hydrogen production. Renewable & Sustainable Energy Reviews 2007;11 (3): 401-425.

[16] Galinska A, Walendziewski J. Photocatalytic water splitting over Pt-TiO_2 in the presence of sacrificial reagents. Energy & Fuels 2005;19 (3): 1143-1147.

[17] Maeda K, Hashiguchi H, Masuda H, et al. Photocatalytic activity of (Ga1-xZnx)(N1-xOx) for visible-light-driven H-2 and O-2 evolution in the presence of sacrificial reagents. Journal of Physical Chemistry C 2008;112 (9): 3447-3452.

[18] Jing DW, Zhang YJ, Guo LJ. Study on the synthesis of Ni doped mesoporous TiO_2 and its photocatalytic activity for hydrogen evolution in aqueous methanol solution. Chemical Physics Letters 2005;415 (1-3): 74-78.

[19] Jing DW, Guo LJ. A novel method for the preparation of a highly stable and active CdS photocatalyst with a special surface nanostructure. Journal of Physical Chemistry B 2006;110 (23): 11139-11145.

[20] Lin WC, Yang WD, Huang IL, et al. Hydrogen Production from Methanol/Water Photocatalytic Decomposition Using Pt/TiO2-xNx Catalyst. Energy & Fuels 2009;23: 2192-2196.

[21] Dey GR, Nair KNR, Pushpa KK. Photolysis studies on HCOOH and HCOO⁻ in presence of TiO_2 photocatalyst as suspension in aqueous medium. Journal of Natural Gas Chemistry 2009;18 (1): 50-54.

[22] Li Y, Xie Y, Peng S, et al. Photocatalytic hydrogen generation in the presence of chloroacetic acids over Pt/TiO2. Chemosphere 2006;63 (8): 1312-1318.

[23] Li YX, Lu GX, Li SB. Photocatalytic hydrogen generation and decomposition of oxalic acid over platinized TiO2. Applied Catalysis a-General 2001;214 (2): 179-185.

[24] Patsoura A, Kondarides D, Verykios X. Photocatalytic degradation of organic pollutants with simultaneous production of hydrogen. Catalysis Today 2007;124 (3-4): 94-102.

[25] Chen XF, Gu WT, Wang SJ, et al. GC/MS Analysis of the Products from Supercritical Methanolysis of Rice-stalk Powder. Energy Sources Part a-Recovery Utilization and Environmental Effects 2010;32 (9): 809-817.

[26] Comini E, Faglia G, Sberveglieri G, et al. Sensitivity enhancement towards ethanol and methanol of TiO2 films doped with Pt and Nb. Sensors and Actuators B-Chemical 2000;64 (1-3): 169-174.

[27] Du GH, Chen Q, Che RC, et al. Preparation and structure analysis of titanium oxide nanotubes. Applied Physics Letters 2001;79 (22): 3702-3704.

[28] Xing C, Jing D, Liu M, et al. Photocatalytic hydrogen production over $Na_2Ti_2O_4(OH)_2$ nanotube sensitized by CdS nanoparticles. Materials Research Bulletin 2009;44 (2): 442-445.

[29] Chen YB, Wang LZ, Lu GQ, et al. Nanoparticles enwrapped with nanotubes: A unique architecture of CdS/titanate nanotubes for efficient photocatalytic hydrogen production from water. Journal of Materials Chemistry 2011;21 (13): 5134-5141.

[30] Li Y, Guo L, Zhang X, et al. Hydrogen production from coal gasification in

supercritical water with a continuous flowing system. International Journal of Hydrogen Energy 2010;35 (7): 3036-3045.

[31] Chen J, Lu Y, Guo L, et al. Hydrogen production by biomass gasification in supercritical water using concentrated solar energy: System development and proof of concept. International Journal of Hydrogen Energy 2010;35 (13): 7134-7141.

[32] Hao X, Guo L, Zhang X, et al. Hydrogen production from catalytic gasification of cellulose in supercritical water. Chemical Engineering Journal 2005;110 (1-3): 57-65.

[33] Buhler N, Meier K, Reber JF. Photochemical hydrogen production with cadmium sulfide suspensions. Journal of Physical Chemistry 1984;88 (15): 3261-3268.

[34] Ma RZ, Bando Y, Sasaki T. Nanotubes of lepidocrocite titanates. Chemical Physics Letters 2003;380 (5-6): 577-582.

[35] Kukovecz A, Hodos M, Konya Z, et al. Complex-assisted one-step synthesis of ion-exchangeable titanate nanotubes decorated with CdS nanoparticles. Chemical Physics Letters 2005;411 (4-6): 445-449.

[36] Zhang SL, Zhou JF, Zhang ZJ, et al. Morphological structure and physicochemical properties of nanotube TiO2. Chinese Science Bulletin 2000;45 (16): 1533-1536.

[37] Jing D, Liu M, Shi J, et al. Hydrogen production under visible light by photocatalytic reforming of glucose over an oxide solid solution photocatalyst. Catalysis Communications 2010;12 (4): 264-267.

[38] Fu X, Long J, Wang X, et al. Photocatalytic reforming of biomass: A systematic study of hydrogen evolution from glucose solution. International Journal of Hydrogen Energy 2008;33 (22): 6484-6491.

RECENT ADVANCES IN MEMBRANE DEVELOPMENT FOR CO_2 FREE FOSSIL POWER PLANTS

Tim Van Gestel, Stefan Baumann, Mariya Ivanova, Wilhelm Meulenberg, Hans Peter Buchkremer
Forschungszentrum Jülich GmbH, Institute of Energy and Climate Research, IEK-1: Materials Synthesis and Processing, Leo-Brandt-Strasse, D-52425 Jülich, Germany

ABSTRACT

This overview paper focuses on the actual developments in ceramic gas separation membranes in our lab, in particular membranes in which a selective transport for H_2/CO_2 and O_2/N_2 gas mixtures can be achieved. Depending on the kind of power plant, different separation tasks exists. Air separation (O_2/N_2) is the focus of the oxyfuel process. In the pre-combustion process an additional H_2/CO_2 separation is included. Here, the state of the art in the development of three classes of membranes is reported: (1) dense O_2-transport membranes; (2) dense H_2-transport membranes; (3) microporous H_2/CO_2-separation membranes. For H_2/CO_2 separation, microporous amorphous SiO_2 membranes have been considered due to their promising combined values of selectivity and flux. A sol-gel dip-coating method is used to prepare the membranes, which results in the deposition of ultra-thin microporous layers (~100 nm) with an average selectivity > 30 and the required H_2 flux. As an alternative, dense H_2 transport membranes are also currently investigated. These membranes are made of crystalline $Ln_{6-x}WO_{12-\delta}$ H^+-conducting materials and exhibit an infinite separation factor. The current concern related to these materials is the low H_2 flux, which requires research on the development of dense thin film layers. In our lab, an inverse tape-casting process has been developed, which currently enables the formation of graded membranes with a 30 μm thick gas-tight $Ln_{6-x}WO_{12-\delta}$ layer. For the oxyfuel process, O_2 transport membranes are investigated. Our current membranes are made of a perovskite material (LSCF, BSCF) by a similar inverse tape-casting process. These membranes have also an infinite separation factor and by lowering the thickness to approximately 20 μm, a flux exceeding 10 ml/min.cm² has been achieved. Research is currently directed towards the development of membranes with a thickness of ~1 μm using a nanoparticle deposition process. Long-term stability tests are under investigation for each membrane type.

INTRODUCTION

1. Power plant processes

A large source of CO_2 emission in the atmosphere is today the energy production with fossil power plants, with coal-fired plants being the main contributor. Since, there are no readily available alternatives, fossil power plants will probably be the dominant source of electrical power the complete century. Therefore, a lot of research is focused on modifying existing plants and plans are made to construct new large-scale CO_2-free plants.

Concepts for the reduction of CO_2 are generally divided into three categories: (1) postcombustion, (2) oxyfuel combustion, (3) precombustion. The first concept is to capture the CO_2 from flue gas after the combustion process and then the cleaned gas, which contains primarily water vapour and N_2, can be released into the atmosphere. This concept is of interest since it can be practiced in existing plants. In fact, some plants have already a CO_2 separation unit, in which the CO_2 capture is based on solvent absorption.

Postcombustion capture has however major disadvantages including the energy requirements for solvent regeneration, large volumes of solvents are needed and expired solvent needs to be disposed. A membrane unit which can remove selectively CO_2 from N_2 should therefore be viewed as

115

a major solution. Organic polymeric membranes are already in an advanced stage of development and are generally considered for the postcombustion concept, which involves the separation of CO_2 and N_2 at temperatures below 100°C. For an excellent review on polymeric membranes for CO_2/N_2 separation, the reader is referred to the review of Powell and Qiao [1]. They present different classes of polymeric membranes and a number of copolymers and blends. The CO_2/N_2 separation factor is typically in the range 20–40, but measurements are usually performed at a temperature of 35°C or lower.

The second concept is the oxyfuel concept. In order to avoid said problems with the separation of CO_2 from N_2 in the flue gas, fossil fuel is burned in nearly pure oxygen rather than in air. This produces a N_2-free flue gas which contains mainly CO_2 and water. By cooling the flue gas, a pure CO_2 stream can then rather easily be obtained. The separation problem in this concept is the separation of O_2/N_2 before the burner (air separation). With current technologies this requires also a high amount of energy and therefore research is focused on the development of efficient O_2/N_2 separation devices, for example a membrane unit. The desired properties of a membrane for this application would include the ability to withstand operating temperatures in the range 750–900°C and operating pressures in the range 20–30 bar. Further, the membrane should have a high O_2 permeability. Major drawbacks for the oxyfuel concept are the high operation temperature, which has, in the case of a membrane unit, important consequences regarding sealing and housing.

In the precombustion concept, the carbon content is removed from the fuel before burning it. The first step is to convert the fuel into a synthesis gas that primarily consists of CO and H_2. Next, the CO is reacted with steam in a shift reactor to produce CO_2. The CO_2 is separated and the remaining H_2-rich gas is combusted in a gas turbine, which results in an exhaust gas that consists of water. Often, this technology is considered as a technology of choice for new-build power plants. Advantages are that multiple fuels can be used and the plant can be used for the production electricity, hydrogen and chemicals. A disadvantage in comparison with the former concepts is the need to construct new power plants. Another significant disadvantage is the use of a solvent absorption column, although the separation of CO_2 from H_2 before the combustion process is easier than the separation of CO_2 from N_2 afterwards and allows the use of smaller absorption columns. However, important gains may also come here from the use of a membrane unit. For more information on the described power plant concepts, the reader is also referred to a few excellent references [2-6].

For the oxyfuel and precombustion concept, with typical membrane operation conditions at a pressure exceeding 20 bar and a temperature in the range 750–900 °C and 200–400°C, respectively, inorganic membranes are usually proposed. Although it is difficult to determine which membrane material will be the best choice (e.g. metallic, ceramic, hybride, …), ceramic membranes are considered today by a number of research groups. Main attractions of ceramic membranes are the cost-efficient preparation methods, which meet the low cost required by commercialization, and the enhanced mechanical, thermal and chemical resistance of ceramic materials in comparison with other membrane materials.

In our group, membrane research has been focused to date mainly on three classes of ceramic membrane materials: (1) dense O^{2-}-conducting materials, (2) dense H^+-conducting materials and (3) microporous materials. The goals of these research efforts are straightforward and can be briefly summarized as follows: low-cost and effective membrane preparation method; a high O_2 permeation, for oxyfuel application; a high H_2 permeation and H_2/CO_2 selectivity for pre-combustion application; resistance to thermal and chemical degradation; mechanical resistance/robustness

MEMBRANE DEVELOPMENT

1. O^{2-}-conducting membranes for O_2/N_2-separation in the oxyfuel concept

Appropriate materials for O_2-separation from air are mixed ionic-electronic conductors (MIEC). Mostly, perovskites and fluorites are considered as membrane materials [7]. Perovskites show typically the highest permeability – a well-known example is $Ba_{0.5}Sr_{0.5}Co_{0.8}Fe_{0.2}O_{3-\delta}$ (BSCF) [8] – but perovskites with a high permeability show also typically a poor chemical stability e.g. in atmosphere containing CO_2, SO_2, or H_2O and particularly in reducing conditions.

The high oxygen permeation in perovskites is due to the conduction mechanism of oxygen ions diffusing through oxygen vacancies in the crystal lattice, which occurs by hopping of the next neighboured oxygen ion into a vacancy. According to the so-called Wagner equation, the oxygen permeation through the membrane is inversely proportional to the membrane thickness and therefore the membrane should be made as thin as possible, without affecting its gas-tightness and mechanical robustness.

In our lab, an inverse tape-casting process has been developed in order to prepare graded membranes with a thin gas-tight functional layer. Figures 1a and 1b show examples of two BSCF membranes, synthesized with the tape-casting process, whereby the functional layer is cast firstly and then the supporting layers are cast subsequently. Basically, the slurries for both casting steps have the same composition and in the second step a pore former is added. Here, the addition of 30wt% corn starch gives a porosity of 40% for the support after the firing process. In this step, the blade gap is set at 1.9 mm, which gives a support thickness of 1 mm after firing. Finally, samples with the desired size are punched out of the green tape and fired for 5h at 1100°C.

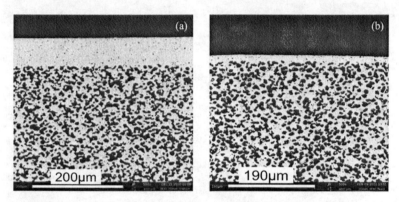

Figure 1. Graded $Ba_{0.5}Sr_{0.5}Co_{0.8}Fe_{0.2}O_{3-\delta}$ membranes made by an inverse tape-casting process.

Currently, research is focussed on reducing the thickness of the functional layer by reducing the doctor blade gap to 100 μm, 50 μm and even smaller. As shown in Figure 1, the layer thickness has been reduced in this way already significantly from around 100 μm in the first attempts (1a) to the range 20–30 μm (1b).

Figure 2 shows the results of oxygen permeation tests with the membrane shown in Figure 1b, with air and pure oxygen as feed gases, respectively. With still a comparably large thickness of 20–30 μm, fluxes of ~20 ml/min.cm² and 4 ml/min.cm² have been obtained at 850°C, with pure oxygen and air as feed gases, respectively, which confirms the large potential of the developed membrane assembly [9].

Research is currently directed towards the development of 5–10 μm thick membranes using a micro-tape-casting process and nanoparticle deposition methods (e.g. spin-coating, dip-coating) are investigated for the synthesis of layers with a thickness of ~1 μm. It should however also be mentioned that below a characteristic thickness, the surface exchange reactions between oxygen molecules and the membrane become rate limiting, which requires an acceleration of the surface exchange kinetics by catalytic activation [9]. Long-term stability investigation of the membranes is in the experimental stage.

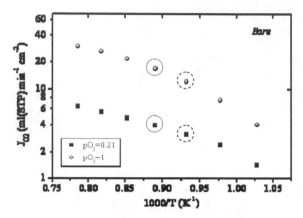

Figure 2. O_2-permeation of the membrane shown in Figure 1b. Measurement with air (black squares) and with pure O_2 (grey circles) as the feed gas. The marked values were measured at 800°C (stripes) and 850°C (dots).

Another group of materials which are under consideration for the synthesis of O_2-separation membranes are Ce-based fluorites, e.g. materials with composition $Ce_{1-x}B_xO_{2-\delta}$ (B= Gd,Pr,Tb). In these materials, the electronic conductivity is rate limiting for oxygen diffusion as it is typically orders of magnitudes lower than the ionic contribution [7,10]. Thus, for practical application the electronic conductivity needs to be enhanced. Advantages of doped ceria membranes include the lower activation energy, which is favourable for operating in lower temperature range, and the chemical stability, which is significantly higher than that of perovskites. Particularly, for application in reducing atmosphere, ceria membranes are considered as a viable alternative for perovskite membranes.

Since the permeation cannot be expected to be as high as this one of the best performing perovskites, the development of thin film membranes is essential. In our lab, we are currently considering two preparation methods to obtain such thin films, with a thickness in the micrometer range. Figure 3 shows examples of gadolinea-doped ceria thin film membrane layers, prepared by spin-coating of nanoparticles (Fig. 3a) and physical vapour deposition (PVD) (Fig. 3b). The main challenge is here the achievement of gas-tight layers and until now the best results in terms of reproducibility are obtained using a spin-coating process, which gives dense layers with a thickness of 1–2 μm and a leak rate for air in the range $5.10^{-4} - 3.10^{-3}$ mbar.l.s^{-1}.cm^{-2}. Coating experiments with physical vapour deposition (PVD) resulted in the formation of somewhat thinner layers (ca. 1 μm), with a gas-tightness in the range $7.10^{-4} - 8.10^{-3}$ mbar.l.s^{-1}.cm^{-2} (measured with He). From the first experiments, it appears that the application of additional bias power during the sputtering process is helpful in order to obtain a dense structure.

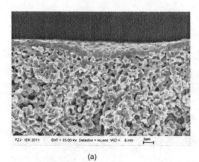 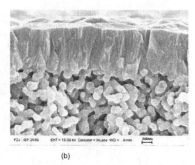

(a)   (b)

Figure 3. SEM pictures of gadolinea-doped ceria thin film layers (20% Gd). (a) Prepared by spin-coating and sintering at 1400°C (5h); (b) Prepared with physical vapour deposition (magnetron sputtering), 400W bias power.

2. H^+-conducting membranes for H_2/CO_2 separation in the precombustion concept

For the separation of H_2 from CO_2 in the precombustion concept, dense and microporous membranes are currently investigated in our lab [11,12].

Dense membranes with proton conductivity are attracting attention since they exhibit in analogy with dense oxygen conducting membranes an intrinsically infinite separation factor. The first extensively researched group of materials include cerates of Ba and Sr (e.g. $SrCeO_3$, $BaCeO_3$, $SrZrO_3$). The main drawback of cerate-based membranes, despite the high proton conduction they exhibit, is their instability under strongly reducing atmosphere in the presence of water vapour and at high operating temperatures. Zirconate-based membranes show a higher stability under reducing atmosphere, but experimental data about the H_2-flux through such membranes is rather limited.

Currently, lanthanoide tungstates are increasingly studied as a membrane material for H_2-separation membranes, due to their improved stability under the relevant operation conditions (e.g. power plant). The feasibility of tungstates in membrane application by comparing them to the state of the art 5% Yb-doped $SrCeO_3$ was estimated by Haugsrud et al. [13]. Based on the ambipolar proton-electron conductivity which is twice higher than the conductivity of perovskite at 1000 °C it is suggested that undoped La_6WO_{12} is a good candidate for membrane applications. Moreover, the behavior of Gd_6WO_{12} and Er_6WO_{12} is pointed out as interesting.

In our lab, membranes based on La_6WO_{12} are currently also under investigation. The hydrogen flux measurements of the first La_6WO_{12} (bulk) membranes were recently reported by Meulenberg et al. [10]. In Figure 4, these measurements are shown and compared with the hydrogen flux through a Nd_6WO_{12} (bulk) membrane with a thickness of 510 μm, recently reported by Escolástico et al. [14]. The membrane made in our lab is a pressed disk with a thickness of 1 mm and a diameter of 25 mm and was measured at Sintef (Oslo, Norway). At the feed side, a mixture of moist H_2/He was used and at the sweep side Ar (both sides at atmospheric pressure). As shown in Figure 4, the hydrogen content has a clear effect on the flux according to the Wagner equation, with a significant increase in the flux at higher hydrogen content. However, basically for these bulk membranes very low fluxes <0.01 ml/min.cm² have been obtained, at a comparably high operation temperature of 800°C for a precombustion application.

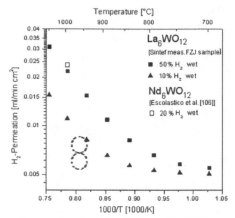

Figure 4. Hydrogen permeation flux of La_6WO_{12} and Nd_6WO_{12} membrane as a function of temperature. The marked values were measured at 800°C. Taken with permission, from reference Nr. 10.

According to the Wagner equation, the permeation through the membrane should also be inversely proportional to the membrane layer thickness. Therefore, an effective way to improve the hydrogen permeation of the mixed proton-electron conducting membrane is to prepare a graded membrane, consisting of a dense functional toplayer and a porous substrate prepared from the same proton-conducting material.

Currently, the main research topic in our lab is the development of such graded membranes by tape-casting and micro-tape-casting, in a similar way as shown in the previous part for oxygen conducting membranes. Figure 5a and 5b show an example of a La_6WO_{12} membrane with a thickness of ~30–40 µm, which has been made with a similar inverse tape-casting process as reported for the oxygen conducting membranes. Summarized, the first step in this process includes casting the functional layer with a blade gap of 100 µm, using a slurry with a particle size of ca. 0.7 µm. In the second step, the support is casted using a slurry of the same powder with the addition of 25wt% of rice starch, which acts as a pore former. The blade gap is set at 1.5 mm in this casting step. Finally, samples with the desired size are punched out and fired for 12h at 1500°C.

The final objective is the development of thin functional layers with a thickness of 10 µm or even lower. For a layer thickness of 10 µm, Norby et al. [15] communicated a permeability data prediction for La_6WO_{12}, $SrCeO_3$ and Er_6WO_{12} as membrane materials. The flux was calculated considering a feed side pressure of 10 atm H_2 and assuming that the flux was ruled by Wagner equation, i.e. through the bulk diffusion mechanism. The predicted hydrogen permeation for La_6WO_{12} at 800°C has a value of about 2.0 ml_n/cm^2min, which is higher compared to the values for $SrCeO_3$ and Er_6WO_{12} at the same temperature of about 1.0 ml_n/cm^2min and less than 0,1 ml_n/cm^2min, respectively.

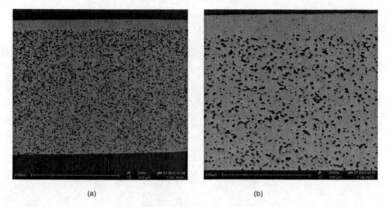

(a)   (b)

Figure 5. Graded La_6WO_{12} membrane, made by an inverse tape-casting process, with a toplayer thickness of ~30–40 μm.

3. Microporous membranes for H_2/CO_2 separation in the precombustion concept

Another class of membranes which are under investigation for H_2/CO_2 separation include microporous membranes. A microporous separation membrane is a graded multilayer porous ceramic membrane – with macro-, meso- and microporous layers – in which the last membrane layer shows a pore size < 2 nm. According to IUPAC notation, microporous materials have a pore diameter < 2 nm, mesoporous materials have pore diameters between 2 nm and 50 nm and macroporous materials have a pore diameter > 50 nm [16]. Microporous membranes are considered by many research groups, because they hold the potential to combine the required selectivity with a high gas permeation. The main classes of microporous membranes include amorphous SiO_2 and doped SiO_2 membranes, templated SiO_2 membranes and zeolite membranes.

For H_2/CO_2 separation, amorphous (doped) SiO_2 membranes have been the most extensively investigated. Amorphous SiO_2 membranes are basically synthesized through two methods: sol-gel coating and chemical vapor deposition (CVD). In analogy with the previous membrane types, the development of thin film functional layers is also here one of the main issues. An important advantage of this class of membranes is the availibility of a large number of well-known sol-gel processing routes, which can be used to deposit layers with a thickness as small as 100 nm [17]. Also, for the preparation of certain mesoporous sublayers, several sol-gel processing routes are available in the literature (e.g. for alumina [18,19]).

In Table 1, typical gas permeation results for a series of membranes, prepared in our lab (cleanroom class 1000) with a sol-gel dip-coating process are listed. The membranes consist of a thin film microporous amorphous SiO_2 toplayer with a thickness of ~100 nm and mesoporous γ-Al_2O_3 sublayers. The mesoporous layers were made from sols containing γ-alumina colloidal particles with a size of ~ 30 nm. The sol preparation was based on the well-known Yoldas process, which includes hydrolysis of a metal-organic precursor ($Al(OC_4H_9)_3$, Sigma-Aldrich) with H_2O and subsequent destruction of larger agglomerates with HNO_3 at elevated temperature (> 80 °C) [19]. The SiO_2 sol was synthesized with a well-known method described by e.g. Uhlhorn [20] and de Lange [21,22], starting from a $Si(OC_2H_5)_4$ precursor in ethanol. Dip-coating was carried out with an automatic dip-coating device, equipped with a holder (vacuum chuck) for square or disk-shaped substrates. In the dip-coating process, sol particles are deposited as a membrane film by contacting the upper-side of the substrate with the coating liquid, while an under-pressure is applied at the back-side. SiO_2 toplayers were fired in air at 400°C for 2 h and a double coating – calcination cycle was used. Intermediate mesoporous γ-Al_2O_3 layers were fired at 600°C.

The observed results with an average H_2/CO_2 selectivity of 30 were in fact in line with expectations. Similar membranes reported in the literature show a similar behaviour and have yielded also a relatively high selectivity. Further, it appears also that a negligible permeation is obtained for N_2 (in 8 out of 10 samples), which gives a very high selectivity for H_2 towards N_2. Currently, we are further optimizing all parameters in the preparation process (e.g. support, sol, coating) in order to further improve the reproducibility of the membrane.

Table 1. Reproducibility tests: Gas permeation of 2 series of 5 membranes with SiO_2 toplayers in l/h.m².bar

Membrane	ΔP (bar)	He (l/h.m².bar)	H_2 (l/h.m².bar)	CO_2 (l/h.m².bar)	I_2 (l/h.m².bar)	H_2/CO_2
γ-Al_2O_3 mesoporous sublayer	0.65	8084	11273	2566	3175	4.39
SiO_2-S11 500°C	4	1970	259	4,1	1.3	63.2
SiO_2-S21 500°C	4	1553	346	12.2	2.3	28.4
SiO_2-S31 500°C	4	1902	233	1.9	0.8	122.6
SiO_2-S41 500°C	3.5	1784	381	4	0	95.2
SiO_2-S51 500°C	2.5	2704	1053	164	10.5	6.4
SiO_2-S12 500°C	4	1564	173	4.2	2.5	41.2
SiO_2-S22 500°C	4	996	140	0.9	0.9	156.1
SiO_2-S32 500°C	4	1018	143	0.5	0.8	286
SiO_2-S42 500°C	4	1348	305	11.6	1.15	26.3
SiO_2-S52 500°C	3	2074	1221	44	19	27.7

S11, S21, S31, ... : Sample 1 of series 1, Sample 2 of series 1, Sample 3 of series 1, ...

As shown in Figure 6a, the gas permeation properties of the membranes with SiO_2 toplayers were also in agreement with the micropore diffusion model (Permeation He > H_2 > CO_2 > N_2). The origin of the selectivity can be described in terms of the smaller kinetic diameter of He and H_2 in comparison with CO_2 and N_2. It appears that these smaller gases can pass through a network of cavities which are formed by tetraeder SiO_4^{4-} units, while larger gas molecules are excluded or in other words that the material functions as a kind of 'molecular sieve'. This hypothesis is also supported by the significantly larger permeability observed for the smallest gas He. For a more detailed description of the gas transport mechanism in silica thin films, the reader is referred to reference Nr.23.

An important drawback mentioned frequently for amorphous SiO_2 membranes is a degradation of the membrane material in a water (vapour) containing atmosphere at a high temperature and pressure. For this reason, a number of researchers focused on developing alternative SiO_2 materials with an improved stability. One common approach is to improve the material stability by adding doping compounds, such as ZrO_2, TiO_2, Al_2O_3, NiO. An extension of the doped silica membrane work is the synthesis of metal doped membranes (e.g. references 24-26).

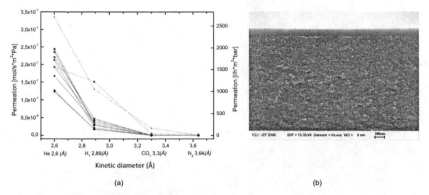

(a)   (b)

Figure 6a. Gas permeation as a function of the kinetic diameter of the test gas for a series of 10 membranes with an amorphous SiO_2 toplayer. Figure 6b. SEM micrograph of a graded membrane with a mesoporous γ-Al_2O_3 sublayer and an amorphous SiO_2 toplayer (bar = 200 nm).

From the viewpoint of thermal and chemical stability, the preparation of crystalline microporous SiO_2 based membranes (e.g. zeolites) is an interesting approach (e.g. references 27-30). Some of these membranes exhibit fascinating properties including a well-defined pore structure with a pore size of approximately 0.5 nm and are not susceptible to densification at higher temperatures or degradation in steam. Several research groups succeeded to prepare zeolite membranes with a good selectivity for some gases, but the existence of intercrystalline pores is often reported as a factor that limits the separation efficiency for small gas molecules.

CONCLUSION

In our research, we consider the development of membrane separation units for new coal power plants, based on the two emerging concepts, pre-combustion and oxyfuel-combustion. The emphasis is on ceramic membranes, in which a selective gas transport for H_2/CO_2 and O_2/N_2 gas mixtures can be achieved. The membranes can basically be classified into two categories: dense and porous membranes.

Initially, the low permeability across the (bulk) dense membranes limited their application as compared to (graded) porous membranes. As shown in this paper, the development of graded membranes with a dense thin film functional layer is an effective way to improve significantly the gas permeation. The current state of the art for dense O_2-transport membrane is very promising. A flux exceeding 10 ml/min.cm² has been found in tape-casted BSCF membranes, with a comparably large thickness exceeding 10 μm. Long-term stability investigation of these membranes is in the experimental stage.

Dense H_2-transport membranes made of crystalline H^+-conducting materials, e.g materials from the series $Ln_{6-x}WO_{12-\delta}$, are considered since they also exhibit an intrinsically infinite separation factor. To achieve a H_2 flux, the current generation of membranes require however a comparably high operational temperature for a pre-combustion application (above 800°C). Therefore, the major challenge exists currently in developing membranes with the smallest possible layer thickness (< 10 μm), e.g. by a micro-tape-casting or sol-gel coating process.

The fabrication of graded porous membranes with thin film functional layers for H_2/CO_2 separation is mature (on lab-scale). These membranes are made by coating thin layers (thickness ~100 nm) via a sol-gel processing route on a porous support material. In our research, it was confirmed that amorphous SiO_2 is a viable material for the synthesis of H_2/CO_2 selective membranes. The average selectivity was around 30 and the reproducibility of the preparation route is 80%.

REFERENCES

[1] C. Powell, G. Qiao, Polymeric CO_2/N_2 gas separation membranes for the capture of carbon dioxide from power plant flue gases, J. Membr. Sci. 279 (2006) 1-49

[2] H.J. Herzog, D. Golomb, Carbon Capture and Storage from Fossil Fuel Use, in C.J. Cleveland (ed.), Encyclopedia of Energy, Elsevier Science Inc., New York, p. 277-287 (2004)

[3] Rune Bredesen, Kristin Jordal, Olav Bolland, High-temperature membranes in power generation with CO_2 capture, Chemical Engineering and Processing 43 (2004) 1129-1158

[4] J. Marano, J. Ciferino, Integration of Gas Separation Membranes with IGCC - Identifying the right membrane for the right job, Energy Procedia 1 (2009) 361-368

[5] J.W. Phair, S.P.S. Badwal, Materials for separation membranes in hydrogen and oxygen production and future power generation, Science and Technology of Advanced Materials 7 (2006) 792-805

[6] H. Yang, Z. Xu, M. Fan, R. Gupta, R. Slimane, A. Bland, I. Wright, Progress in carbon dioxide separation and capture: A review, J. Environmental Sci. 20(2008) 14-27

[7] J. Sunarso, S. Baumann, J.M. Serra, W.A. Meulenberg, S. Liu, Y.S. Lin, J.C. Diniz da Costa. Mixed ionic–electronic conducting (MIEC) ceramic-based membranes for oxygen separation. J. Membr. Sci. 320 (2008) 13–41

[8] Z. Shao, W. Yang, Y. Cong, H. Dong, J. Tong, G. Xiong. Investigation of the permeation behavior and stability of a $Ba_{0.5}Sr_{0.5}Co_{0.8}Fe_{0.2}O_{3-\delta}$ oxygen membrane. J. Membr. Sci. 172 (2000) 177–188

[9] S. Baumann, J.M. Serra, M.P. Lobera, S. Escolástico, F. Schulze-Küppers, W.A. Meulenberg, Ultrahigh oxygen permeation flux through supported $Ba_{0.5}Sr_{0.5}Co_{0.8}Fe_{0.2}O_{3-\delta}$ membranes, J. Membr. Sci. 377 (2011) 198–205

[10] W.A. Meulenberg, I. Voigt, R. Kriegel, S. Baumann, M. Ivanova, T. Van Gestel, Inorganic Membranes for CO_2 Separation", in "Efficient Carbon Capture Coal Power Plants – Process Engineering for CCS Power Plants", Wiley-VCH Verlag, Weinheim, Edited by Detlef Stolten and Victor Scherer, ISBN: 978-3-527-33002-7, (2011), p. 319-350

[11] W.A. Meulenberg, M. E. Ivanova, T. Van Gestel, M. Bram, H.P. Buchkremer, D. Stöver, J.M. Serra, State of the art of ceramic membranes for hydrogen separation", in Hydrogen and Fuel Cells, Wiley-VCH Verlag, Weinheim, Edited by Detlef Stolten, ISBN: 978-3-527-32711-9, (2010) p. 321-349

[12] M. Bram, T. Van Gestel, W.A. Meulenberg, Nano-Structured Ceramic Membranes for carbon Capture and Storage CCS", in Nanotechnology and Energy, Chapter 3: Examples for nanotechnological applications in the energy sector. Editor: Prof. Voß, in press, (2012)

[13] R. Haugsrud, Ch. Kjølseth, Effects of protons and acceptor substitution on the electrical conductivity of La_6WO_{12}. J. Phys. Chem. Sol. 69 (2008) 1758–1765.

[14] S. Escolástico, V.B. Vert, J.M. Serra, Preparation and characterization of nanocrystalline mixed proton–electronic conducting materials based on the system Ln_6WO_{12}, Chem. Mater. 2009, 21, 3079–3089.

[15] T. Norby, R. Haugsrud, High Temperature Proton Conducting Materials for H_2-separation. Invited lecture at FZJ Workshop, November 16, 2007

[16] http://www.iupac.org/goldbook/M03853.pdf

[17] C.J. Brinker, G.W. Scherer, Sol-Gel Science: The Physics and Chemistry of Sol-Gel Processing, Academic Press, Inc.

[18] V.T. Zaspalis, W. Van Praag, K. Keizer, J.R.H. Ross and A.J. Burggraaf, Synthesis and characterization of primary alumina, titania and binary membranes, J. Mat. Sci. 27 (1992) 1023-1035

[19] B.E. Yoldas, Preparation of glasses and ceramics from metal-organic compounds, J. Mat. Sci. 12 (1977) 1203-1208

[20] R. J. R. Uhlhorn, K. Keizer, A. J. Burggraaf, Gas transport and separation with ceramic membranes. Part II. Synthesis and separation properties of microporous membranes J. Membr. Sci. 66 (1992) 271-287

[21] R.S.A. de Lange, J.H.A. Hekkink, K. Keizer, A.J. Burggraaf, Polymeric-silica-based sols for membrane modification applications: sol-gel synthesis and characterization with SAXS, J. Non-Cryst. Solids 191 (1995) 1-16

[22] R. S. A. de Lange, J. H. A. Hekkink, K. Keizer, A. J. Burggraaf, Formation and characterization of supported microporous ceramic membranes prepared by sol-gel modification techniques, J. Membr. Sci. 99 (1995) 57-75

[23] S.T. Oyama, D. Lee, P. Hacarlioglu, R.F. Saraf, Theory of hydrogen permeability in nonporous silica membranes, J. Membr. Sci. 244 (2004) 45-53

[24] Y. Gu, P. Harcarlioglu, S.T. Oyama, Hydrothermally stable silica-alumina composite membranes for hydrogen separation, J. Membr. Sci. 310 (2008) 28-37

[25] M. Kanezashi, M. Asaeda, Hydrogen permeation characteristics and stability of Ni-doped silica membranes in steam at high temperature, J. Membr. Sci. 271 (2006) 286-93

[26] S. Battersby, T. Tasaki, S. Smart, B. Ladewig, S. Liu, M.C. Duke, V. Rudolph, J.C. Diniz da Costa, Performance of cobalt silica membranes in gas mixture separation, J. Membr. Sci. 329 (2009) 91-98

[27] M. Hong, S. Li, J.L. Falconer, R.D. Noble, Hydrogen purification using a SAPO-34 membrane, J. Membr. Sci. 307 (2008) 277-283

[28] V. Sebastian, I. Kumakiri, R. Bredesen, M. Menendez, Zeolite membrane for CO$_2$ removal: Operating at high pressure, J. Membr. Sci. 292 (2007) 92-97

[29] F. Bonhomme, M.E. Welk, T.M. Nenoff, CO2 selectivity and lifetimes of high silica ZSM-5 membranes, Micr. Mesop. Mat. 66 (2003) 181-18

[30] N. Nishiyama, M. Yamaguchi, T. Katayama, Y. Hirota, M. Miyamoto, Y. Egashira, K. Ueyama, K. Nakanishi, T. Ohta, A. Mizusawa, T. Satoh, Hydrogen-permeable membranes composed of zeolite nano-blocks, J. Membr. Sci. 306 (2007) 349-354

ELECTROSPINNING OF NANOCOMPOSITE SCAFFOLDS FOR BONE TISSUE ENGINEERING: EMITTING ELECTRODE POLARITY AND CHARGE RETENTION

Ho-Wang Tong, Min Wang
Dept. of Mechanical Engineering, The University of Hong Kong, Pokfulam Road, Hong Kong
* Email: memwang@hku.hk

ABSTRACT

   Electrospinning is a simple and versatile technique for producing ultrafine fibers for many applications including tissue engineering. Most electrospinning studies conducted by researchers around the world have been using high positive voltages [i.e., positive voltage electrospinning (PVES)] and negative voltage electrospinning (NVES) has been rarely investigated. There was virtually no report on NVES of tissue engineering scaffolds until recently. However, fibrous scaffolds made by NVES, which can retain negative charges, may elicit desirable cell responses than fibrous scaffolds made by PVES in terms of cell attachment, proliferation and differentiation. The current study investigated the formation and characteristics of fibrous poly(hydroxybutyrate-co-hydroxyvalerate) (PHBV) polymer scaffolds and fibrous carbonated hydroxyapatite (CHA)/PHBV nanocomposite scaffolds under NVES. PVES of PHBV and CHA/PHBV scaffolds was also performed for comparisons. For fabricating fibrous CHA/PHBV nanocomposite scaffolds, CHA nanospheres were first synthesized using a nanoemulsion process and the nanospheres were dispersed in a PHBV solution under ultrasonication for electrospinning. The scaffolds obtained were characterized in terms of fiber diameter, morphology and microstructure and scaffold wettability, charge bearing ability and charge retention. For the biological study, cell culture experiments were performed using a human osteoblastic-like cell line (SaOS-2). It was found that CHA nanospheres were evenly distributed in and along CHA/PHBV nanocomposite fibers. Compared with PHBV polymer scaffolds, CHA/PHBV nanocomposite scaffolds exhibited better wettability. Both types of scaffolds retained detectable residual charges for considerable times and possessed similar charge decay profiles. Although the scaffolds facilitated the proliferation and spreading of SaOS-2 cells, CHA/PHBV nanocomposite scaffolds resulted in much higher expression of alkaline phosphatase (ALP) activity of SaOS-2 cells than PHBV polymer scaffolds.

INTRODUCTION

   Tissue engineering aims to develop biological substitutes for the repair, restoration or regeneration of damaged or diseased body tissues. Compared with the therapy of direct cell injection to the tissue site, the use of scaffolds in tissue engineering has many advantages such as the provision of temporary support for host or implanted cells and modulation of the dimension, shape, strength and composition of the graft *in vitro*. An ideal scaffold should possess all required properties such as high porosity to enhance mass transport, sufficient strength to withstand the shearing forces during cultivation in the bioreactor, and suitable degradation rate that match as much as possible with the rate of neo-tissue formation. There are many scaffold fabrication techniques for tissue engineering and there is a great interest in the construction of nanofibrous scaffolds owing to their distinctive properties such as biomimicry of the structure and function of extracellular matrix (ECM) of most human body tissues and extremely high surface-area-to-volume ratios. The methods for producing nanofibrous tissue engineering scaffolds include vapor-phase polymerization, extraction, drawing, template synthesis, phase separation, self-assembly, and electrospinning [1].
   Electrospinning, a technique capable of forming nanofibrous scaffolds from a great variety of materials with different structures and properties, has received considerable attention in the biomedical field since more than 10 year ago [2]. While most reported electrospinning studies were conducted

using high positive voltages [i.e., positive voltage electrospinning (PVES)], only very few researchers investigated negative voltage electrospinning (NVES) and the scarcity of research on NVES has been pointed out in several review articles [3-5]. There was virtually no report on NVES of tissue engineering scaffolds until our publication appeared recently. One major difference between PVES and NVES is that the former provides positively charged nanofibers while the latter produces negatively charged nanofibers. It is well known that biomaterial surfaces bearing positive or negative charges can significantly influence cell-material interactions [6-11]. Therefore, an ability to control the surface charge will be potentially valuable for achieving required cell responses. It was reported that for some polymers, long-term charge retention in electrospun fibers could occur after electrospinning [12, 13], which may have significant implications in tissue engineering.

Poly(hydroxybutyrate-*co*-hydroxyvalerate) (PHBV) is a natural and biodegradable polymer and has been successfully electrospun into fibrous tissue engineering scaffolds in our previous studies [14]. But pure PHBV polymer scaffolds lack osteoconductivity which is desired for bone tissue engineering. The osteoconductivity of a scaffold can be enhanced when the scaffold contains a substantial amount of bioactive bioceramics such as hydroxyapatite (HA) [15]. Therefore, the incorporation of HA particles into electrospun fibers to construct fibrous HA/polymer composite scaffolds has received considerable attention [16]. Carbonated HA (CHA) is a proven osteoconductive and biodegradable bioceramic. CHA nanoparticles, which mimic bone apatite more chemically and exhibit a higher resorption rate than HA nanoparticles, are more desirable than pure HA for bone regeneration. Incorporating CHA nanoparticles in the scaffolds could lead to the formation of nanocomposite scaffolds having better osteoconductivity and biodegradability for bone tissue engineering. From the scaffold fabrication point of view, it is easier to incorporate spherical bioceramic nanoparticles in composite scaffolds than using nanoparticles with angular shapes. Therefore, in common with many other researchers, most of our investigations have involved spherical CHA nanoparticles (i.e., CHA nanospheres).

The aim of this study was to investigate the formation of fibrous PHBV polymer and CHA/PHBV nanocomposite scaffolds via PVES and NVES and evaluate the scaffolds in different areas including the charge bearing ability and biological properties.

MATERIALS AND METHODS

Fabrication of PHBV and CHA/PHBV scaffolds via PVES or NVES

PHBV with an average molecular weight of 222.2 kDa and 5mol% of hydroxyvalerate was obtained from Sigma-Aldrich, USA. The solvent for making polymer solutions, chloroform (analytical grade), was supplied by VWR BDH Prolabo, UK. The polymer and solvent were used in the as-received state. CHA nanospheres were synthesized in-house using a nanoemulsion process which was similar to the process established in our previous study [17].

To fabricate pure PHBV polymer scaffolds, PHBV was dissolved in chloroform at a polymer concentration of 15% w/v and the solution was subjected to PVES or NVES using the facility described previously [18]. To investigate the best way to construct CHA/PHBV nanocomposite scaffolds, three different routes were employed for fabricating CHA/PHBV nanocomposite fibers having different CHA contents (5, 10 and 15wt%). Route I was a two-step process which involved mixing CHA nanospheres with a PHBV solution before electrospinning. Route II was also a two-step process and involved the application of ultrasonic energy to the CHA/PHBV mixture by an ultrasonic cell disruptor (Branson Sonifier® Model 450D, USA) for dispersing CHA nanospheres before electrospinning. Route III was a one-step process which involved the direct application of ultrasonic energy to the CHA/PHBV mixture during electrospinning. It was found in PVES that among the three routes, Route II was the only promising approach for fabricating usable CHA/PHBV nanocomposite scaffolds. Therefore, nanocomposite scaffolds fabricated via Route II, under either PVES or NVES, were used for subsequent experiments.

Characterization of PHBV and CHA/PHBV scaffolds fabricated via PVES or NVES

PHBV and CHA/PHBV scaffolds fabricated via PVES or NVES were examined using a scanning electron microscope (SEM, Hitachi S-3400 N, Japan). The fiber diameters were measured (n = 50) by analyzing each SEM micrograph using an image analysis tool (UTHSCSA Image Tool, USA). The presence of CHA nanospheres in CHA/PHBV nanocomposite fibers was studied using spot analysis of energy-dispersive X-ray spectroscopy (EDX) and the distribution of these nanospheres in nanocomposite scaffolds was evaluated using EDX elemental mapping. The wettability of each type of fibrous scaffold (n = 3) was assessed using a contact angle measuring machine (SL200B, Shanghai Solon Tech Inc. Ltd, China) and deionized water as the test liquid. The electrical potential, which reflected the amount of retained charges in electrospun scaffolds, was measured directly at the surface of each fibrous scaffold using an electrostatic voltmeter with an accuracy of 0.1V (IsoProbe, Model 244A, USA). The first time point for electrical potential measurement was 1 minute after PVES or NVES while the subsequent time points were 1, 2, 3, 4, 5, 6, 7, 14, 21, 28, 60 and 90 days. The *in vitro* biological evaluation was performed according to our established methodology and used a human osteoblast-like cell line (SaOS-2) [19]. The cell morphology, cell proliferation and expression of alkaline phosphatase (ALP) activity were assessed during the cell culture period of 14 days. The cell morphology was observed under SEM after the cells with the scaffold had been fixed and processed by a critical point drier. The cell proliferation was assessed using 3-(4,5-dimethylthiazol-2-yl)-2,5-diphenyltetrazolium bromide (MTT) assay while the ALP activity was determined by measuring the amount of p-nitrophenol (pNP) formation per milligram of protein per hour.

RESULTS

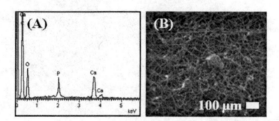

Figure 1. Energy-dispersive X-ray spectroscopy (EDX) analysis of electrospun CHA/PHBV nanocomposite scaffolds: (A) EDX spot analysis; (B) EDX mapping. (Red spots: CHA nanospheres)

Route I could construct CHA/PHBV nanocomposite scaffolds with the CHA contents ranging from 5wt% to 15wt%. However, CHA nanospheres were not homogeneously distributed along the fibers and they often agglomerated as CHA clusters. When the CHA content became 15wt%, the metal needle for electrospinning was easily blocked during the electrospinning process and the fiber yield was very low. Route II was also found to be capable of constructing CHA/PHBV nanocomposite scaffolds with CHA contents up to 15wt%. Figure 1 shows the EDX results for CHA/PHBV nanocomposite scaffolds fabricated via Route II. The Ca and P peaks in the EDX spectrum in Figure 1 confirmed the presence of CHA nanospheres in nanocomposite fibers while the EDX mapping shown as Figure 1(B) illustrated the homogeneous distribution of CHA nanospheres in the whole nanocomposite scaffold, though some nanospheres still agglomerated as small clusters. It was found that Route III led to extremely quick evaporation of the solvent and hence rapid solidification of the solution during electrospinning, which hindered the continuity of the fiber formation process and resulted in insufficient fibers for scaffold formation.

Figure 2 shows fibrous PHBV polymer and CHA/PHBV nanocomposite scaffolds fabricated via PVES or NVES. The average diameter of PHBV fibers electrospun under NVES or PVES was about 3 μm and the average diameter of CHA/PHBV fibers was slightly larger. The surface of CHA/PHBV nanocomposite fibers was rougher than that of the PHBV polymer fibers. The polarity of the applied voltage did not appear to significantly affect the fiber diameter and morphology. The water contact angles of PHBV polymer scaffolds fabricated via PVES and NVES were 104.2 ± 3.1° and 104.8 ± 3.6°, respectively. However, the water contact angle of CHA/PHBV nanocomposite scaffolds was not measurable no matter whether PVES or NVES was employed (Figure 3).

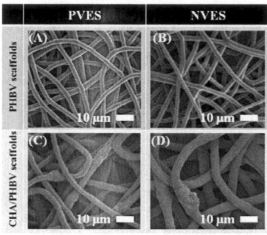

Figure 2. PHBV polymer and CHA/PHBV nanocomposite scaffolds fabricated via PVES or NVES: (A, B) PHBV polymer scaffolds; (C, D) CHA/PHBV nanocomposite scaffolds; (A, C) scaffolds fabricated via PVES; (B, D) scaffolds fabricated via NVES.

The initial surface potential of PHBV polymer scaffolds fabricated via PVES, CHA/PHBV nanocomposite scaffolds fabricated via PVES, PHBV polymer scaffolds fabricated via NVES and CHA/PHBV nanocomposite scaffolds fabricated via NVES was 79.5 ± 3.6 V, 77.7 ± 3.3 V, -86.6 ± 4.1 V and -88.7 ± 4.5 V, respectively. All scaffolds exhibited similar charge decay profiles and retained charges for over 20 days after electrospinning. The residual charges on these scaffolds became non-detectable after 60 days of electrospinning (Figure 4).

Figure 5 displays the morphology of SaOS-2 cells seeded on PHBV polymer scaffolds and CHA/PHBV nanocomposite scaffolds after 7 days of cell culture. Both PHBV and CHA/PHBV scaffolds retained their fibrous architecture during the whole cell culture period, even though they had been immersed in Dulbecco's Modified Eagle Medium at 37°C for up to 14 days. The cells could not only attach to but also expanded and spread well in all directions on both types of fibrous scaffolds. Numerous filopodia were observed around the cells after 14 days of cell culture. Figure 6 shows the results of SaOS-2 cell proliferation on tissue culture polystyrene plate (the control) and two types of

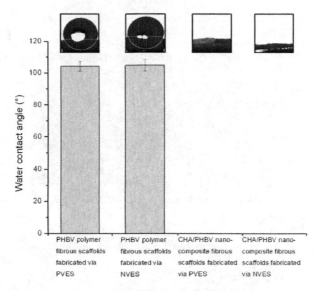

Figure 3. Water contact angle of PHBV polymer and CHA/PHBV nanocomposite scaffolds fabricated via PVES or NVES. (Top row: side-view of the water droplet on each type of scaffolds.)

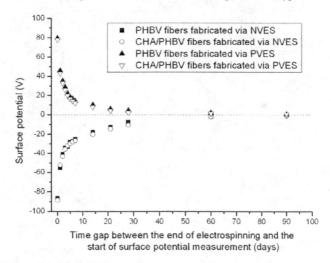

Figure 4. Charge retention of PHBV polymer and CHA/PHBV nanocomposite scaffolds fabricated via PVES or NVES.

fibrous scaffolds (PHBV and CHA/PHBV) using MTT assay. The cells seeded on both types of scaffolds proliferated significantly from day 2 to day 7 of cell culture but did not change significantly from day 7 to day 14. Although the ALP activity of SaOS-2 cells seeded on both types of fibrous scaffolds was comparable to each other on day 2 and day 7 of cell culture, the ALP activity expressed by the cells on CHA/PHBV nanocomposite scaffolds was significantly higher than those on PHBV polymer scaffolds after 14 days of cell culture.

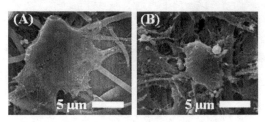

Figure 5. Morphology of SaOS-2 cells after 7 days of cell culture on (A) PHBV polymer scaffolds; (B) CHA/PHBV nanocomposite scaffolds.

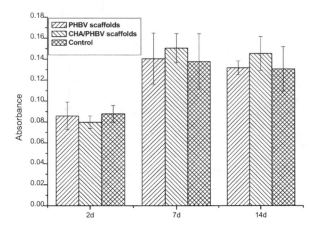

Figure 6. Proliferation of SaOS-2 cells seeded on tissue culture polystyrene plate (the control) and two types of fibrous scaffolds (PHBV and CHA/PHBV) as a function of cell culture time.

DISCUSSION

The method to disperse bioceramic nanoparticles in a polymer solution prior to and/or during electrospinning is very important for forming usable bioceramic/polymer nanocomposite scaffolds. In the current investigation, for Route I, it is believed that CHA nanospheres agglomerated immediately after mixing the nanospheres with the PHBV solution, leading to poor dispersion of CHA nanospheres and hence easy agglomeration of the nanospheres during electrospinning. In Route II, ultrasonic power was applied to the CHA/PHBV mixture during the solution preparation process such that the CHA nanospheres were highly dispersed in the mixture. The CHA/PHBV solution containing dispersed CHA nanospheres was immediately subjected to electrospinning, resulting in the formation of nanocomposite fibers having reasonably homogeneous distribution of CHA nanospheres. The advantage of this method (i.e., using Route II) is that the nanospheres could be distributed in composite fibers reasonably well without the use of any dispersant, thus minimizing any biocompatibility concern. However, some CHA nanospheres still agglomerated owing to the time gap between the solution preparation process and the electrospinning process. To eliminate this time gap, Route III was investigated. In Route III, CHA nanospheres should not agglomerate during the whole fiber formation process because the particle-polymer solution was subjected to the continuous application of ultrasonic power. But the rapid evaporation of solvent from the polymer solution in Route III undermined entirely its effectiveness for nanocomposite fiber formation and insufficient nanocomposite fibers were generated before the electrospinning process came to a complete stop. Due to the intrinsic limitations of Route I and Route III, Route II appeared to be the only viable method to fabricate usable nanocomposite fibrous scaffolds for bone tissue engineering.

The fiber diameter was not varied significantly by the incorporation of CHA nanospheres in fibers or by the change of polarity of the applied voltage. It was found in our previous studies that the ultimate fiber diameter depended mainly on the viscosity, conductivity and surface tension of the electrospinning solution. When these factors were not changed by sufficient degrees, the fiber diameter would not be changed significantly. As the CHA nanospheres could not only be encapsulated inside nanocomposite fibers but also attached on the fiber surface, the increased roughness of the nanocomposite fibers was basically due to the presence of CHA nanospheres on or near the fiber surface. The CHA nanospheres on the fiber surface also enhanced the wettability of electrospun scaffolds. It is known that cells tend to adhere and spread on hydrophilic surfaces rather than on hydrophobic surfaces. The use of fibrous scaffolds with surface CHA nanospheres would enhance osteoblastic cell adhesion and proliferation, making the scaffolds good candidates for bone tissue engineering.

In the current investigation, the major difference was that PVES produced positively charged fibers and NVES resulted in negatively charged fibers. Immediately after PVES or NVES, a lot of residual charges were likely to reside on the surface of fibers. The randomly oriented fibers in electrospun scaffolds prevented most of the fibers from making contact with the conductive fiber collector and hence minimized the dissipation of charges. After electrospinning, the surface charges were neutralized continuously by ions in the air, leading to the charge decay. The surface potential detected after a period of time, e.g., 60 days, could be due to the entrapped charges within the nanofibrous scaffolds. A recent study revealed that the residual charges on electrospun fibers came from two sources: residual surface charges and excess trapped charges in fibers [20]. Upon soaking electrospun fibrous scaffolds in deionized water or phosphate buffered saline (PBS), residual surface charges are exposed to the aqueous medium and they may be dissipated. However, for semicrystalline polymers, electric charges can be trapped at the crystalline/amorphous interface [21]. It was shown that electrospun poly(butylene terephthalate) fibers possessed a high level of residual charge trapped in or near the crystalline phase as a result of charge injection during the electrospinning process [20]. As these trapped electric charges are not exposed directly to deionized water or PBS, the may not be dissipated easily. It was noted in another study that when electrospun fibers were plasma coated with a film with electric surface charges, the electric charges on fibers were not dissipated in a biological system and these charges affected the biological response [22].

The incorporation of bioceramics, such as HA and tricalcium phosphate (TCP) into electrospun fibers has been investigated in recent years and some promising results were reported. Zhou *et al.* fabricated carbonated calcium deficient hydroxyapatite (CDHA)/poly(lactic-acid) (PLA) nanocomposite fibers via electrospinning and found that the incorporation of CDHA decreased PLA fiber diameters, accelerated PLA degradation, buffered pH decrease caused by PLA degradation, and improved the bioactivity and biocompatibility of the scaffold [23]. Combing melt-plotting and electrospinning, Yeo *et al.* produced a new hierarchical scaffold consisting of a melt-plotted TCP/polycaprolactone (PCL) composite and embedded collagen nanofibers. Compared with pure polymer scaffolds, the hierarchical composite scaffolds showed significant water absorption ability, enhanced wettability, better mechanical properties, and higher proliferation rate of osteoblast-like cells (MG63 cells) [24]. But these studies on bioceramic/polymer nanocomposite fibers via electrospinning have been focusing on PVES. The formation of bioceramic/polymer nanocomposite fibers via NVES could render the fibrous scaffolds with distinctive properties and our investigation has gained the first insights into NVES of fibrous nanocomposite scaffolds for tissue engineering.

It has been widely reported that the behavior of cells seeded on biomaterials can be affected by the polarity and intensity of electrical charges on the biomaterials [6-11]. Some studies have particularly showed the desirable cell behaviors on negatively charged substrates. Tarafder *et al.* constructed biphasic calcium phosphate composites and found that negatively charged surfaces elicited better osteoblast responses than positively charged or neutral surfaces in terms of cell attachment, cell proliferation and ECM formation [25]. Wang *et al.* fabricated chitosan membranes through tape casting and found that the viability of osteoblasts and the ALP activity expressed by osteoblasts seeded on negatively charged surfaces were higher than those seeded on uncharged surfaces [26]. Our successful formation of CHA/PHBV nanocomposite fibrous scaffolds via NVES may provide distinctive scaffolds for bone tissue engineering. For *in vitro* biological evaluation, it was found in the current investigation that after the rapid cell proliferation from day 2 to day 7, the number of proliferated cells on both types of fibrous scaffolds did not change significantly. Since cellular proliferation decreases when cells differentiate into mature, ECM-producing cells [27], the cells might have begun the process of differentiation just like other osteoblastic cells would naturally do. This cellular activity was advantageous because it allowed for the generation of critical ECM proteins, leading to new bone tissue formation. It is also worth noting that ALP synthesis is one of the major steps within the whole differentiation sequence of bone cells. Therefore, the high ALP activity of SaOS-2 cells cultured on CHA/PHBV nanocomposite scaffolds also clearly indicated that the CHA/PHBV nanocomposite scaffolds are a suitable candidate for bone regeneration.

CONCLUSIONS

Fibrous PHBV polymer and CHA/PHBV nanocomposite scaffolds were formed via PVES or NVES. Ultrasonication before electrospinning was an effective method for dispersing CHA nanospheres in polymer solutions, leading to the formation of fibrous CHA/PHBV nanocomposite scaffolds with homogeneous distribution of CHA nanospheres. The incorporation of CHA nanospheres into fibers did not significantly affect the fiber diameter but greatly enhanced the wettability of scaffolds. PVES produced positively charged PHBV polymer fibers and CHA/PHBV nanocomposite fibers and the fibers fabricated via NVES were negatively charged. The amount of retained charges on PHBV polymer and CHA/PHBV nanocomposite fibrous scaffolds decreased over time after PVES or NVES. Compared with PHBV polymer scaffolds, CHA/PHBV nanocomposite scaffolds elicited higher ALP activity expressed by the cells. Varying scaffold composition and retained charge provides important means for influencing the behavior of cells coming into contact with the scaffolds.

ACKNOWLEDGEMENTS
This work was supported by GRF grants (HKU 7176/08E and HKU 7181/09E) from the Research Grants Council of Hong Kong. The authors thank Professor William Lu and technical staff in the Department of Orthopaedics and Traumatology, HKU, for assistance with the cell culture work.

REFERENCES

1.   V. Beachley and X. Wen, Polymer nanofibrous structures: fabrication, biofunctionalization, and cell interactions, *Progress in Polymer Science*, **35**, 868-92 (2010).
2.   M.P. Prabhakaran, L. Ghasemi-Mobarakeh and S. Ramakrishna, Electrospun composite nanofibers for tissue regeneration, *Journal of Nanoscience and Nanotechnology*, **11**, 3039-57 (2011).
3.   A. Greiner and J.H. Wendorff, Electrospinning: a fascinating method for the preparation of ultrathin fibers, *Angewandte Chemie*, **46**, 5670-703 (2007).
4.   V.E. Kalayci, P.K. Patra, Y.K. Kim, S.C. Ugbolue and S.B. Warner, Charge consequences in electrospun polyacrylonitrile (PAN) nanofibers, *Polymer*, **46**, 7191-200 (2005).
5.   W.-E. Teo, R. Inai and S. Ramakrishna, Technological advances in electrospinning of nanofibers, *Science and Technology of Advanced Materials*, **12**, 013002 (2011).
6.   Y.M. Chen, J.P. Gong, M. Tanaka, K. Yasuda, S. Yamamoto, M. Shimomura and Y. Osada, Tuning of cell proliferation on tough gels by critical charge effect, *Journal of Biomedical Materials Research Part A*, **88A**, 74-83 (2009).
7.   T.-H. Chung, S.-H. Wu, M. Yao, C.-W. Lu, Y.-S. Lin, Y. Hung, C.-Y. Mou, Y.-C. Chen and D.-M. Huang, The effect of surface charge on the uptake and biological function of mesoporous silica nanoparticles in 3T3-L1 cells and human mesenchymal stem cells, *Biomaterials*, **28**, 2959-66 (2007).
8.   N.J. Hallab, K.J. Bundy, K. O'Connor, R. Clark and R.L. Moses, Cell adhesion to biomaterials: correlations between surface charge, surface roughness, adsorbed protein, and cell morphology, *Journal of Long-Term Effects of Medical Implants*, **5**, 209-31 (1995).
9.   J.A. Hunt, B.F. Flanagan, P.J. McLaughlin, I. Strickland and D.F. Williams, Effect of biomaterial surface charge on the inflammatory response: evaluation of cellular infiltration and TNF alpha production, *Journal of Biomedical Materials Research*, **31**, 139-44 (1996).
10.   M. Ohgaki, T. Kizuki, M. Katsura and K. Yamashita, Manipulation of selective cell adhesion and growth by surface charges of electrically polarized hydroxyapatite, *Journal of Biomedical Materials Research*, **57**, 366-73 (2001).
11.   G.B. Schneider, A. English, M. Abraham, R. Zaharias, C. Stanford and J. Keller, The effect of hydrogel charge density on cell attachment, *Biomaterials*, **25**, 3023-28 (2004).
12.   D. Lovera, C. Bilbao, P. Schreier, L. Kador, H.-W. Schmidt and V. Altstädt, Charge storage of electrospun fiber mats of poly(phenylene ether)/polystyrene blends, *Polymer Engineering & Science*, **49**, 2430-39 (2009).
13.   H.-W. Tong and M. Wang, Negative voltage electrospinning and positive voltage electrospinning of tissue engineering scaffolds: a comparative study and charge retention on scaffolds, *Nano LIFE*, **2**, 1250004 (2012).
14.   H.W. Tong, M. Wang and W.W. Lu, Electrospun poly(hydroxybutyrate-*co*-hydroxyvalerate) fibrous membranes consisting of parallel-aligned fibers or cross-aligned fibers: characterization and biological evaluation, *Journal of Biomaterials Science -- Polymer Edition*, **22**, 2475-97 (2011).
15.   M. Wang, Developing bioactive composite materials for tissue replacement, *Biomaterials*, **24**, 2133-51 (2003).
16.   J. Venugopal, M.P. Prabhakaran, Y. Zhang, S. Low, A.T. Choon and S. Ramakrishna, Biomimetic hydroxyapatite-containing composite nanofibrous substrates for bone tissue engineering, *Philosophical Transactions of the Royal Society A: Mathematical, Physical and Engineering Sciences*, **368**, 2065-81 (2010).

17. W.Y. Zhou, M. Wang, W.L. Cheung, B.C. Guo and D.M. Jia, Synthesis of carbonated hydroxyapatite nanospheres through nanoemulsion, *Journal of Materials Science: Materials in Medicine*, **19**, 103-10 (2008).
18. H.W. Tong and M. Wang, Electrospinning of fibrous polymer scaffolds using positive voltage or negative voltage: a comparative study, *Biomedical Materials*, **5**, 054110 (2010).
19. H.W. Tong, M. Wang, Z. Li and W.W. Lu, Electrospinning, characterization and *in vitro* biological evaluation of nanocomposite fibers containing carbonated hydroxyapatite nanoparticles, *Biomedical Materials*, **5**, 054111 (2010).
20. L.H. Catalani, G. Collins and M. Jaffe, Evidence for Molecular Orientation and Residual Charge in the Electrospinning of Poly(butylene terephthalate) ļanofibers, *Macromolecules*, **40**, 1693-97 (2007).
21. Y. Arita, S. Sha Shiratori and K. Ikezaki, A method for detection and visualization of charge trapping sites in amorphous parts in crystalline polymers, *Journal of Electrostatics*, **57**, 263-71 (2003).
22. J.E. Sanders, S.E. Lamont, A. Karchin, S.L. Golledge and B.D. Ratner, Fibro-porous meshes made from polyurethane micro-fibers: Effects of surface charge on tissue response, *Biomaterials*, **26**, 813-18 (2005).
23. H. Zhou, A.H. Touny and S.B. Bhaduri, Fabrication of novel PLA/CDHA bionanocomposite fibers for tissue engineering applications via electrospinning, *Journal of Materials Science: Materials in Medicine*, **22**, 1183-93 (2011).
24. M. Yeo, H. Lee and G. Kim, Three-dimensional hierarchical composite scaffolds consisting of polycaprolactone, -tricalcium phosphate, and collagen nanofibers: Fabrication, physical properties, and in vitro cell activity for bone tissue regeneration, *Biomacromolecules*, **12**, 502-10 (2011).
25. S. Tarafder, S. Bodhak, A. Bandyopadhyay and S. Bose, Effect of electrical polarization and composition of biphasic calcium phosphates on early stage osteoblast interactions, *Journal of Biomedical Materials Research Part B: Applied Biomaterials*, **97B**, 306-14 (2011).
26. Y.Y. Wang, Y.L. Qu, P. Gong, P. Wang, Y. Man and J.D. Li, Preparation and in vitro evaluation of chitosan bioelectret membranes for guided bone regeneration, *Journal of Bioactive and Compatible Polymers*, **25**, 622-33 (2010).
27. G.S. Stein, J.B. Lian and T.A. Owen, Bone cell differentiation: A functionally coupled relationship between expression of cell-growth- and tissue-specific genes, *Current Opinion in Cell Biology*, **2**, 1018-27 (1990).

GRAPHITIC CARBON NITRIDE MODIFIED BY SILICON FOR IMPROVED VISIBLE-LIGHT-DRIVEN PHOTOCATALYTIC HYDROGEN PRODUCTION

Po Wu, Jinwen Shi, Jie Chen, Bin Wang, Liejin Guo[*]
International Research Center for "Solar-Hydrogen" Renewable and Clean Energy, State Key Laboratory of Multiphase Flow in Power Engineering, Xi'an Jiaotong University, 28 West Xianning Street, Xi'an 710049, PR China.

ABSTRACT

In this work, composites of silica and graphitic carbon nitride (g-C$_3$N$_4$) were conveniently synthesized through polymerization of melamine and ethyl silicate. Some basic physicochemical properties of these metal-free photocatalysts were characterized by X-ray diffraction (XRD), X-ray photoelectron spectrum (XPS), Fourier transform infrared spectrometer (FTIR) spectra, transmission electron microscope (TEM), UV-visible diffuse reflectance spectrum and N$_2$ adsorption-desorption measurement. The XRD patterns of all the prepared materials were dominated by the characteristic (002) peak at 27.5° of a graphitic structure, indicating that the induced Si did not destroy the crystal structure of g-C$_3$N$_4$. As the initial molar ratio of Si to C increased, enhanced photocatalytic activity was observed for H$_2$ evolution from a triethanolamine aqueous solution under visible-light irradiation over the series of composites. It was proved that the introduction of Si resulted in large surface area, which is favorable for reactant-transfer and charge-migration, beneficial for photocatalysis.

KEYWORDS: Graphitic carbon nitride, Photocatalytic hydrogen generation, Porosity.

INTRODUCTION

In the past few decades, photocatalytic hydrogen production from water over semiconductor catalysts under sunlight has attracted increasing attention, due to the global energy and environment issues[1-5]. Among the water splitting systems, heterogeneous powder system is believed to enable large-scale application[6]. To date, enormous efforts have been focused on the design of photocatalysts that should be stable, inexpensive, responsive to visible light, as well as efficient for carrier migration. Among the reported materials (e.g., metal oxides, nitrides, sulfides, phosphides, organic-metal complexes), sulfides (e.g., CdS, Cd$_x$Zn$_{1-x}$S) have shown excellent activities for hydrogen generation from water in the presence of Na$_2$S/Na$_2$SO$_3$[7, 8]. However, the toxicity and photocorrosion severely restrict their practical application.

---

[*] *Corresponding author:*   E-mail address: lj-guo@mail.xjtu.edu.cn

Recently, graphitic carbon nitride (g-C_3N_4) with an optical band gap of 2.7 eV, has shown promising performance as a metal-free photocatalyst for hydrogen evolution and organics oxidation under visible light irradiation[9-11]. This material is considered to be stable under light irradiation in water solution as well as in acidic or basic solution due to the strong covalent bonds between carbon and nitride atoms[12]. Moreover, modifications (e.g., doping, composite constructing) have been extensively investigated to improve the photocatalytic activity of pure g-C_3N_4. It is reported that the hydrogen generation rate of the hybrid material 10%-Zn/g-C_3N_4 was more than 10 times higher than that of pure g-C_3N_4[13]. Sulfur-doped g-C_3N_4 with a unique electronic structure had an activity 8 times higher than C_3N_4 under visible light irradiation ($\lambda > 420$ nm)[14]. Interestingly, polymer composites of g-C_3N_4 and poly (3-hecylthiophene) was reported to show a H_2 evolution rate 300 times the yield achieved using g-C_3N_4[12]. In addition, the optical absorption of g-C_3N_4 could be extended up to 750 nm in the visible light region by convenient copolymerization with organic monomers like barbituric acid[15]. Furthermore, it has been studied that enhanced photocatalytic performance of g-C_3N_4 could be achieved by the introduction of porosity through the hard templating approaches[6, 9, 16], due to the favorable transfer of reactants and photogenerated carriers.

In this study, a facile thermal polymerization route was employed to synthesize g-C_3N_4 and the composites of SiO_2 and g-C_3N_4 (the initial atomic ratios of Si/C are 10%, 50%, 100%), with the consideration of charge compensation. The introduction of Si seemed to hinder the illimitable bulk growth of the carbon nitride phase at a nanometer level, leading to large specific surface area and improved photocatalytic hydrogen evolution.

EXPERIMENTAL SECTION

Preparation of Samples

All reagents were of analytical purity and used without further purification. Initially, 5 g melamine was added and stirred in 200 mL water, followed by the addition of 5 drop of HNO_3 and required amount of ethyl silicate (the molar ratios of [ethyl silicate]:[melamine]x3 were 0%, 10%, 50%, 100%). Then, the resulting mixtures were stirred and heated at 90 °C-100 °C to remove water. After that, white powders obtained were heated to 550 °C within 4 h and kept at this temperature for another 4 h. The obtained pure carbon nitride and composites were denoted as g-C_3N_4 and Si-g-C_3N_4, respectively.

Characterization of Materials

The X-ray diffraction (XRD) patterns of photocatalysts were obtained from a X-ray diffractometer (Panalytic X'pert Pro), with Ni-filtered Cu Kα irradiation (wavelength 1.5406 Å). The UV-visible optical absorption spectra were measured on a spectrophotometer (Hitachi UV4100), equipped with a lab-sphere diffuse reflectance accessory. The crystallite morphologic micrograph was observed by a transmission electron microscopy (TEM: JEOL, JEM-2100). N_2 adsorption-desorption isotherms were conducted at 77 K in a surface area and porosity analyzer (Micromeritics ASAP2020). Surface areas and cumulative pore volumes were determined by the Brunauer-Emmett-Teller (BET) and Barret-Joyner-Halenda (BJH) method, respectively. X-ray photoelectron spectrum (XPS) data were obtained on a Kratos spectrometer (AXIS Ultra DLD) with a monochromatic Al Kα line source (150W, hν=1486.69eV). All binding energies were referenced to the C1s peak at 284.3 eV. Fourier transform infrared spectrometer (FTIR) spectra were recorded on a time resolution spectrometer (BRUKER VERTEX 70). Photoluminescence (PL) spectra were obtained by a fluorescence spectrophotometer (Hitachi F-3010).

Evaluation of Photocatalytic Activity

Photocatalytic hydrogen evolution was performed in a gas-closed system with a side irradiation Pyrex cell. Photocatalyst (0.2 g) was dispersed in triethanolamine (TEOA) aqueous solution (160mL H_2O; 20mL TEOA). Platinum (1.0 wt %) as co-catalyst was photodeposited in situ on the photocatalyst from precursors of $H_2PtCl_6 \cdot 6H_2O$. Nitrogen was purged through the cell for 10 min before reactions to remove oxygen. The reaction temperature was maintained at 35±1 °C by thermostatic circulating water. The photocatalysts were irradiated with visible light through a cutoff filter (λ>420nm) from a Xe lamp (PLS-SXE300, 361 mW/cm^2). The concentration of hydrogen evolved was determined with an on-line thermal conductivity detector (TCD) gas chromatography (TDX-01 column). Control experiments showed no appreciable hydrogen evolution without irradiation or photocatalysts.

RESULTS AND DISCUSSION

Crystal Structure

The XRD patterns of the as-synthesized polymeric g-C$_3$N$_4$ and the Si-containing derivatives (Si-g-C$_3$N$_4$) are shown in Figure 1. It can be observed that g-C$_3$N$_4$ and Si- g-C$_3$N$_4$ have similar patterns that are dominant at 27.5° of the typical graphitic interlayer (002) peak and 13.0° of the in-plane structural packing motif (100) peak[9, 17].As the ratio of Si/C increases, the samples show a decrease of relative intensities at the typical g-C$_3$N$_4$ peaks, accompanied by an increased intensity of broad peak packet at 22.4° indicative of amorphous SiO$_2$[18]. It can be assumed that the introduction of Si rendered the formation of a spot of SiO$_2$ and a spatial confinement to the growth of extended g-C$_3$N$_4$. On one hand, this limitation may result in small grain size and large surface area of g-C$_3$N$_4$. On the other hand, it can hinder the weight loss of g-C$_3$N$_4$ by preventing the sublimation of melamine. However, the formation of SiO$_2$ has not destroyed the graphitic structure of the host material.

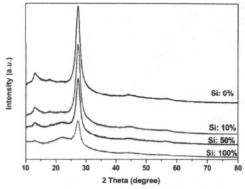

Figure 1. XRD patterns of pure g-C₃N₄ and Si-g-C₃N₄ (with various initial atomic ratio of Si/C)

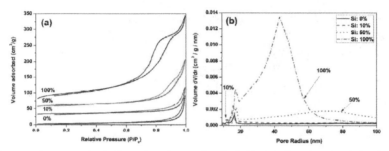

Figure 2. Nitrogen sorption isotherms (a) and pore radius distribution (b) of pure g-C₃N₄ and Si-g-C₃N₄
(with various initial atomic ratio of Si/C)

Morphology

To support the assumption upon the XRD results, Nitrogen adsorption-desorption isotherms of the as-synthesized samples are plotted in Figure 2 (a). As can be seen, all the plots are related to the Type isotherms, indicative of the week interaction between the adsorbents and the adsorbate. At the high relative pressure, a Type H3 hysteresis loop is identified and this type of loop confirms the formation of pores[19]. The pore size distributions of the samples determined from the desorption branch of the isotherms are shown in Figure 2 (b). As is demonstrated, pure g-C₃N₄ has a small quantity of pores with radius less than 20 nm. Compared to this, plentiful large pores appear and increase with the increase of the initial atomic ratio of Si/C. In addition, the BET surface areas and BJH adsorption cumulative pore volumes of the samples have been listed in Table II. From what have been described above, it can be concluded that a large amount of pores were formed due to the confinement of the silicon source, in accordance with the supposition from XRD results. What's more, this can also be explicitly demonstrated by the TEM images (Figure 3). Uniform but unordered pore structures can be observed over the representative Si-g-C₃N₄ (Si/C: 100%).

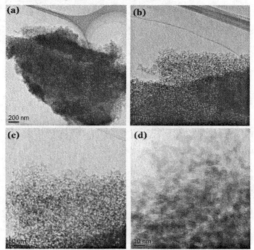

Figure 3. TEM images of Si-g-C$_3$N$_4$ (Si/C: 100%) with different resolution

Optical Properties

At first, the optical properties of the as-synthesized samples can be investigated by the UV-visible diffuse reflectance spectra (Figure 4). As is shown, the absorption edges of all the samples are close to 450 nm, in agreement with the previous reported value of band gap (2.7 eV)[20]. After the introduction of Si, the absorption spectrum did not change much, indicating that the energy band structure of g-C$_3$N$_4$ was unaltered and that SiO$_2$ hardly contribute to the optical absorption.

Then, the FTIR spectra are used to identify the mode of vibration of the samples (Figure 5). In general, the peaks at around 2360 cm^{-1} and 640cm^{-1} are attributed to the vibration of CO$_2$ and cyclic Si-O-Si species[21, 22]. After excluding these peaks, there are three main regions in the spectra: 3140 cm^{-1}, 1240-1640 cm^{-1}, 810 cm^{-1}. A broad band near 3140 cm^{-1} corresponds to the stretching modes of terminal NH$_2$ or NH groups at the edges of the graphene-like CN sheet[23]. Several strong bands in the 1240-1640 cm^{-1} region are the stretching modes of CN heterocyclic[24]. As is well known, the bands at around 810 cm^{-1} is assigned to the characteristic breathing mode of the triazine units. Furthermore, the main absorbance band intensity of Si-g-C$_3$N$_4$ is slightly stronger than that of g-C$_3$N$_4$, indicating the formation of composites of SiO$_2$ and g-C$_3$N$_4$.

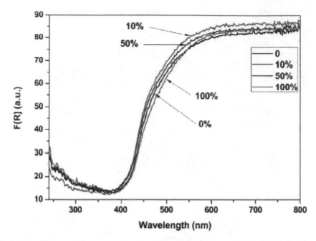

Figure 4. UV-visible diffuse reflectance spectra of pure g-C$_3$N$_4$ and Si-g-C$_3$N$_4$ (with various initial atomic ratio of Si/C)

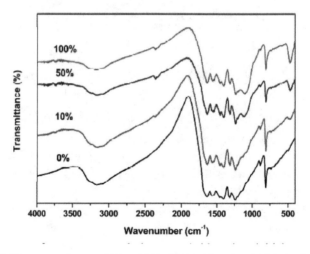

Figure 5. FTIR spectra of pure g-C$_3$N$_4$ and Si-g-C$_3$N$_4$ (with various initial atomic ratio of Si/C)

Table I. Elemental composition of pure g-C_3N_4 and Si-g-C_3N_4 (with various initial atomic ratio of Si/C) obtained from XPS spectra.

Si/C (at.%)	C(at.%)	N(at.%)	O(at.%)	Si(at.%)
0	48.2	48.6	3.2	0
10	36.4	33.5	21.6	8.5
50	29.5	19.9	36.4	14.2
100	24.5	18.6	42.6	14.5

In addition, the chemical bonding of the samples can be studied by XPS spectra (Figure 6) to further confirm the existence of g-C_3N_4 and SiO_2. Pure g-C_3N_4 mainly shows C1s and N1s peaks. In detail, the C1s spectrum can be convoluted into four peaks with binding energies of 293.5, 288.1, 287.8 and 284.5 eV (Figure 6b), the two lowest of which contribute much. The peak at 287.8 eV is assigned to the sp^3-hybridized carbon[9], and 284.5 eV corresponds to pure graphitic sites in the amorphous carbon nitride[9, 25]. Meanwhile, the N1s spectrum shows three peaks centered at 404.3, 399.8 and 398.3 eV (Figure 6c). The main peaks at 399.8 and 398.3 eV correspond to N atoms trigonally bonded to carbon atoms and nitrogen sp^2-bonded to carbon[9], respectively. The above analysis clearly indicates that graphitic carbon nitride was obtained both before and after the introduction of Si. Furthermore, after the incorporation of Si, there are large signals of O and Si located in the XPS spectrum. The peak near 103.5 eV is attributed to Si 2p binding energy in SiO_2, in accordance with the previous reported result[26].

The surface elemental composition of the products can also be obtained from the XPS data (Table I). The overall atomic ratios of C/N and O/Si are larger than 3/4 and 2/1, respectively, due to the moisture, atmospheric O_2 or CO_2 adsorbed on the surface of the samples[9]. Additionally, as the ratio of Si/C increases, the amounts of C and N decrease slightly, accompanied by the increase of percentage of Si and O.

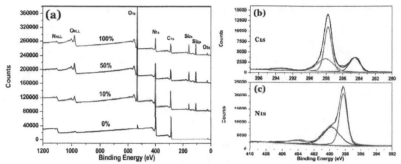

Figure 6. XPS spectra of pure g-C₃N₄ and Si-g-C₃N₄ (with various initial atomic ratio of Si/C): (a) total; (b) C 1s; (c) N1s

Photoluminescence (PL) spectra of the prepared samples were also recorded. As shown in Figure 7, an intense emission peak centered at ca. 470 nm can be observed for all the samples, and this peak intensity decreases with the increase of Si/C ratio. With consideration of the absorption edges (ca. 450 nm) depicted above, the PL peak showed Stoke's shift and could be related to electron-hole recombination processes within the semiconductor. It is thus inferred that the introduction of Si resulted in less recombination, facilitating the carrier transport. This result can be attributed to the increased pores and enlarged surface areas.

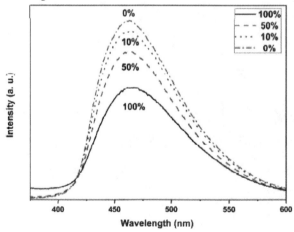

Figure 7. PL spectra of pure g-C₃N₄ and Si-g-C₃N₄ (with various initial atomic ratio of Si/C)

Photocatalytic Activities

The as-synthesized samples g-C$_3$N$_4$ and Si-g-C$_3$N$_4$ show steady hydrogen evolution from water containing TEOA as an electron donor under visible light irradiation (Figure 8). As can be seen from the plotted line, there is a slight increase in the rate of hydrogen generation when more silicon source is added. Compared to pure g-C$_3$N$_4$, Si-g-C$_3$N$_4$ (100%) shows a photocatalytic activity 1.3 times that of the former.

To discuss the mechanism of the enhancement of photocatalytic hydrogen evolution, the surface areas, accumulated pore volumes and hydrogen production rates are listed in Table II. From the table, it can be observed that the rate of hydrogen evolution increases with the increase of surface area and pore volume. Therefore, it is reasonable to claim that the large surface area and pore volume play a crucial role in the improvement of hydrogen production of the as-prepared samples. In detail, it should be noted that SiO$_2$ contained in Si-g-C$_3$N$_4$ didn't have UV-visible optical absorption, approximately invalid to the photocatalytic hydrogen generation. Meanwhile, there was not much alternation of crystal structure and optical properties over g-C$_3$N$_4$. As a result, the enhanced photocatalytic activity relied mainly on the morphology, properly speaking, the large surface area provided by the plenty of pores. These pores could present much transmission channels and reactive sites, of benefit for the diffusion of the reactant and migration of photogenerated electrons and holes. Moreover, if SiO$_2$ is removed by NH$_4$HF$_2$, the porous structure is particularly promising as a host semiconductor support or sensitizer for the design of hybrid visible-light photocatalysts, as it enables the convenient functionalization by deposition or surface reaction[16]. Such work is in progress.

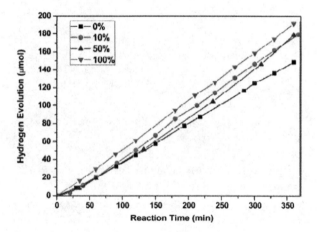

Figure 8. Photocatalytic hydrogen evolution over pure g-C$_3$N$_4$ and Si-g-C$_3$N$_4$ (with various initial atomic ratio of Si/C)

Table II. Porous features and photocatalytic H_2 evolution rate of pure g-C_3N_4 and Si-g-C_3N_4 (with various initial atomic ratio of Si/C)

Si/C (at.%)	Surface area (m^2/g)	Pore volume (cm^3/g)	H_2 evolution $(\mu mol/h/g)$
0	14	0.14	124
10	17	0.14	147
50	43	0.24	149
100	136	0.44	160

CONCLUSION

As an efficient photocatalyst, metal-free composite of g-C_3N_4 and SiO_2 could be successfully prepared by a facile thermal polymerization approach. In the synthetic progress, the introduction of Si did not change much of the crystal structure and optical properties of the host material. Nevertheless, it could hinder the extended growth of g-C_3N_4, ensuring a structure of multi-pores with heavily enlarged surface area and pore volume. This property could facilitate the migration of reactant and photogenerated carriers, leading to enhanced photocatalytic hydrogen production. Furthermore, this work provides an effective method for the synthesis of multi-pore material with large surface area. This material can also serve as support or sensitizer for the design of efficient hybrid visible-light photocatalysts.

ACKNOWLEDGEMENTS

The authors gratefully acknowledge the financial supports from the National Natural Science Foundation of China (No. 50821064) and National Basic Research Program of China (No. 2009CB220000).

REFERENCES

[1] Yi Z, Ye J, Kikugawa N, Kako T, Ouyang S, Stuart-Williams H, et al. An Orthophosphate Semiconductor with Photooxidation Properties Under Visible-Light Irradiation, *J. Nature Material*, **9**, 559-64 (2010).

[2] Kudo A, Miseki Y. Heterogeneous Photocatalyst Materials for Water Splitting, *J. Chemical Society Reviews*, **38**, 253-78 (2009).

[3] Maeda K, Teramura K, Lu D, Takata T, Saito N, Inoue Y, et al. Photocatalyst Releasing Hydrogen From Water, *J. Nature*, **440**, 295- (2006).

[4] Chen XB, Shen SH, Guo LJ, Mao SS. Semiconductor-based Photocatalytic Hydrogen Generation, *J.*

Chemical Reviews, **110**, 6503-70 (2010).

[5] Shen S, Shi J, Guo P, Guo L. Visible-Light-driven Photocatalytic Water Splitting on Nanostructured Semiconducting materials, *J. International Journal of Janotechnology* **8**, 523-91 (2011).

[6] Cui Y, Zhang J, Zhang G, Huang J, Liu P, Antonietti M, et al. Synthesis of Bulk and Nanoporous Carbon Nitride Polymers from Ammonium Thiocyanate for Photocatalytic Hydrogen Evolution, *J. Journal of Materials Chemistry*, **21**, 13032-9 (2011).

[7] Jing D, Guo L. A Novel Method for the Preparation of a Highly Stable and Active CdS Photocatalyst with a Special Surface Nanostructure, *J. The Journal of Physical Chemistry B*, **110**, 11139-45 (2006).

[8] Liu M, Wang L, Lu G, Yao X, Guo L. Twins in $Cd_{1-x}Zn_xS$ Solid Solution: Highly Efficient Photocatalyst for Hydrogen Generation from Water, *J. Energy & Environmental Science*, **4**, 1372-8 (2011).

[9] Vinu A, Ariga K, Mori T, Nakanishi T, Hishita S, Golberg D, et al. Preparation and Characterization of Well-ordered Hexagonal Mesoporous Carbon Nitride, *J. Advanced Materials*, **17**, 1648-52 (2005).

[10] Zhang P, Wang Y, Yao J, Wang C, Yan C, Antonietti M, et al. Visible-light-induced Metal-free Allylic Oxidation Utilizing a Coupled Photocatalytic System of g-C_3N_4 and N-Hydroxy Compounds, *J. Advanced Synthesis & Catalysis*, **353**, 1447-51 (2011).

[11] Wang X, Maeda K, Thomas A, Takanabe K, Xin G, Carlsson JM, et al. A Metal-free Polymeric Photocatalyst for Hydrogen Production From Water Under Visible Light, *J. Jature Material*, **8**, 76-80 (2009).

[12] Yan H, Huang Y. Polymer Composites of Carbon Nitride and Poly(3-hexylthiophene) to Achieve Enhanced Hydrogen Production From Water Under Visible Light, *J. Chemical Communications*, **47**, 4168-70 (2011).

[13] Bing Yue and Qiuye Li and Hideo Iwai and Tetsuya Kako and Jinhua Y. Hydrogen production using zinc-doped carbon nitride catalyst irradiated with visible light, *J. Science and Technology of Advanced Materials*, **12**, 034401 (2011).

[14] Liu G, Niu P, Sun C, Smith SC, Chen Z, Lu GQ, et al. Unique Electronic Structure Induced High Photoreactivity of Sulfur-doped Graphitic C_3N_4, *J. Journal of the American Chemical Society*, **132**, 11642-8 (2010).

[15] Zhang J, Chen X, Takanabe K, Maeda K, Domen K, Epping JD, et al. Synthesis of a Carbon Nitride Structure for Visible-Light Catalysis by Copolymerization, *J. Angewandte Chemie International Edition*, **49**, 441-4 (2010).

[16] Chen X, Jun Y-S, Takanabe K, Maeda K, Domen K, Fu X, et al. Ordered Mesoporous SBA-15 Type Graphitic Carbon Nitride: A Semiconductor Host Structure for Photocatalytic Hydrogen Evolution with Visible Light, *J. Chemistry of Materials*, **21**, 4093-5 (2009).

[17] Ding Z, Chen X, Antonietti M, Wang X. Synthesis of Transition Metal-Modified Carbon Nitride Polymers for Selective Hydrocarbon Oxidation, *J. ChemSusChem*, **4**, 274-81 (2011).

[18] Shi J, Guo L. Photocatalytic Performance of HMS Doped with Chromium or Vanadium for

Hydrogen Production in Aqueous Fomic Acid Solution, *J. ACTA CHIMICA SIĮICA* , **65**, 323-8 (2007).

[19] Sing KSW, Everett DH, Haul RAW, Moscou L, Pierotti RA, Rouquerol J, et al. Reporting Physisorption Data for Gas/Solid Systems. Handbook of Heterogeneous Catalysis: Wiley-VCH Verlag GmbH & Co. KGaA; 2008.

[20] Wang X, Maeda K, Chen X, Takanabe K, Domen K, Hou Y, et al. Polymer Semiconductors for Artificial Photosynthesis: Hydrogen Evolution by Mesoporous Graphitic Carbon Nitride with Visible Light, *J. Journal of the American Chemical Society*, **131**, 1680-1 (2009).

[21] Panda RN, Hsieh MF, Chung RJ, Chin TS. FTIR, XRD, SEM and Solid State NMR Investigations of Carbonate-containing Hydroxyapatite Nano-particles Synthesized by Hydroxide-gel Technique, *J. Journal of Physics and Chemistry of Solids*, **64**, 193-9 (2003).

[22] Primeau N, Vautey C, Langlet M. The Effect of Thermal Annealing on Aerosol-gel Deposited SiO_2 Films: a FTIR Deconvolution Study, *J. Thin Solid Films*, **310**, 47-56 (1997).

[23] Yan H, Yang H. TiO_2-g-C_3N_4 composite materials for photocatalytic H_2 evolution under visible light irradiation, *J. Journal of Alloys and Compounds*, **509**, L26-L9 (2011).

[24] Meng Y, Xin G, Chen D. A Facile Pyrolysis Method for g-C_3N_4 Synthesizing: Photocatalytic Degradation of Methylene Blue Under Visible Light, *J. Optoelectronics and Advanced Materials*, **5**, 648-50 (2011).

[25] Marton D, Boyd KJ, Al-Bayati AH, Todorov SS, Rabalais JW. Carbon Nitride Deposited Using Energetic Species: A Two-Phase System, *J. Physical Review Letters*, **73**, 118-21 (1994).

[26] Stakheev AY, Shpiro ES, Apijok J. XPS and XAES Study of Titania-silica Mixed Oxide System, *J. The Journal of Physical Chemistry*, **97**, 5668-72 (1993).

FABRICATION OF CdS/ZnO CORE–SHELL NANOFIBERS BY ELECTROSPINNING AND THEIR PHOTOCATALYTIC ACTIVITY

Yan Wei[1], Yang Guorui[2]

[1]State Key Laboratory of Multiphase Flow in Power Engineering, Xi'an Jiaotong University, Xi'an, Shaanxi, China.

[2]Department of Environmental Science & Engineering, Xi'an Jiaotong University, Xi'an, Shaanxi, China.

ABSTRACT:

Electrospinning has been proved to be a versatile and effective method for manufacturing micro-scale to nano-scale fibers continuously. In this article, CdS/ZnO core/shell nanofibers were prepared successfully by the typical single-nozzle electrospinning, with subsequent simple thermal decomposition. The SEM and TEM results indicated that core-shell fibers with a diameter of 200-350 nm and shell thickness of 50nm were obtained when PVP/zinc acetate/cadmium acetate/thiocarbamide composite nanofibers were calcined at 480 °C for 4h . The EDS analysis results showed that the core fiber is CdS and the shell layer is ZnO. Hydrogen evolution measurements from photocatalytic water splitting using the ZnO/CdS core-shell nanofibers as photocatalyst were carried out. The ZnO/CdS core-shell nanofibers demonstrate a much higher ability for H_2 evolution than that of sole CdS or ZnO. The highest H_2 evolution rate was up to 354 $\mu molh^{-1}g^{-1}$.

INTRODUCTION

Development of renewable and carbon free energy is one of the most important scientific challenges in the new century, due to the global energy crisis and environmental problems. Hydrogen is an effective energy carrier, which can be able to overcome the disadvantages of using conventional fossil fuels as it has high heat conversion efficiency and zero carbon emission. Since Fujishima and Honda reported the photoelectrochemical water splitting for hydrogen production on a TiO_2 electrode[2], the search for active photocatalyst materials for water splitting has attracted extensive attention. However, the traditional photocatalysts such as TiO_2, ZnO and SnO_2 can only make use of the ultraviolet light which accounts for 5% in the natural light, while do not absorb the visible light which is 43% in the solar spectrum. The development of visible light photocatalysts, therefore, has become an important topic in the photocatalysis research today. CdS is one of the most important visible-light-driven photocatalysts because its relative narrow band gap (2.42 eV) corresponds well with the spectrum of sunlight, and its conduction band edge is more negative than the H_2O/H_2 redox potential. Unfortunately, CdS suffers from photocorrosion and fast recombination of charge carriers. Enormous efforts have been devoted to suppress photocorrosion and increase photoactivity of CdS, among various methods, coupling CdS with another wide-band-gap semiconductor (such as ZnO) seems to be an appealing choice.

As one of the large band-gap semiconductors, Zinc oxide (ZnO), has attracted widespread attention in the photocatalytic domain due to its higher electron mobility than TiO_2, exaction binding energy (60 meV), breakdown strength and exciton stability. In this regard, incorporating CdS and ZnO into an integrated heterostructure is of great interest because the resulting products possess improved physical and chemical properties[3-8]. Many types of CdS/ZnO heterostructure with enhanced photocatalytic activity have been synthesized by different research groups. CdS/ZnO composite particles have been prepared and their photocatalytic activity test showed that CdS/ZnO heterojunctions improved separation efficiency of photoinduced electron/hole pairs, consequently, the higher photocatalytic activity was observed [9]. Youngjo Tak et al.[10] found that the CdS nanoparticles/ZnO nanowires heterostructural arrays which were prepared by a facile two-step solution method had improved photocatalytic activities comparing with the bare ZnO nanowires. Recently, one-dimensional (1D) core/shell-type nanomaterials have received growing attention for their novel properties and potential applications in electronics, catalysis and photonics arising from their unique geometrical structures[11]. Nanorods of CdS/ZnO composite semiconductor were successfully synthsized, and the CdS/ZnO nanorods showed better photocatalytic activity than bare CdS and ZnO under visible light illumination[12]. ZnO/CdS core/shell nanowires sensitized photoelectrochemical cells has a high short-circuit photocurrent density and improved power conversion efficiency. These results demonstrated that ZnO/CdS core/shell nanowires can provide a facile and compatible frame for the potential applications in nanowire-based solar cells[13]. wang et al.[14] reported that the ZnO/CdS core/shell nanorods exhibited stable and high photocatalytic activity for water splitting into hydrogen. The authors suggested that the core/shell structures were in favor of the charge carrier separation and could avoid photocatalyst agglomeration. Paromita Kundu et al.[15] also obtained ZnO/CdS nanoscale heterostructures for highly active photocatalysis under solar irradiation conditions.

Although there has already been much progress and various methods in the synthesis and assembly of ZnO/CdS core/shell heterostructure, effective strategies to produce continuous and tailored 1D core/shell CdS/ZnO nanostructures, which could protect CdS from photocorrosion, are still required. Electrospinning has been proven to be a versatile and simple technique to generate 1D core/shell-type nanofibers, as recent works shown with TiN/VN[16], TiO_2/ZnO[17], Zn-TiO_2/ZnO[18], MoO_2/C[19], CeO_2/TiO_2[20], C/TiO_2[21,22], ZnO/ZnS[23], SnO_2/ZnO[24]. Most of the core/shell nanofibers were prepared by coaxial electrospinning or electrospining combining with other methods including hydrothermal, atomic layer deposition and metal-organic chemical vapor deposition. However, these two strategies suffer from complicated processures and small-scale production. For example, coaxial electrospinning is affected by many factors, such as miscibility or immiscibility of the pair of solutions, viscosity ratio, interfacial tension and so on. As a result, it is difficult to achieve good concentricity of the core and shell materials in the as-electrospun fibers. Using one step typical single-nozzle electrospinng to prepare core/shell nanofibers will be an ideal solution which can avoid the disadvantages as mentioned above.

Against this background, CdS/ZnO core/shell nanofibers were prepared successfully by the typical (single-nozzle) electrospinning with subsequent simple thermal decomposition in this work. The photocatalytic activities of the CdS/ZnO core/shell nanofibers were investigated by the photocatalytic hydrogen generation from water under visible light.

EXPERIMENTAL

Materials:

polyvinylpyrroidone (PVP, Mn=1,300,000) was supplied by ISP. Co., USA. Analytical pure zinc acetate and cadmium acetate were obtained from Tianjing chemical reagent Company. Thiocarbamide and thiourea were obtained from Chengdu chemical reagent Company. Methanol and acetic acid used as solvents were purchased from Tianjing northern chemical Co.

Prepareation of CdS/ZnO core/shell nanofibers:

The preparation of spinning solutions: firstly, zinc acetate, cadmium acetate and thiourea were dissolved thoroughly into the acetic acid / methanol(3:37) mixed solvents; secondly, PVP was added into the above solution. After stirring and aging for about 6 h, the PVP/zinc acetate/cadmium acetate composite gel was obtained. Then the viscous solution of PVP / zinc acetate/cadmium acetate composite gel was filled in a glass syringe. A pinhead, which attaches on the tip of syringe, was connected to a DC high-voltage generator. An aluminum foil with a ceramic tile covered on top served as the counter electrode. The distance between the tip of the pinhead and the aluminum foil was approximately 12 cm. A voltage of 12.0 kV was applied to the solution and a web of fibers was collected on the ceramic tile. The as-spun fibers were finally calcined at 480 °C -700 °C on the ceramic tile for 4 h to induce crystallinity(shown in Fig. 1).

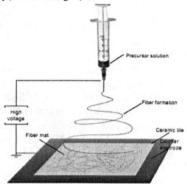

Figure 1. Schematic diagram of the electrospinning process

Preparation of H-CdS:

Firstly, cadmium acetate, thioacetamide were dissolved into 75 ml deionized water, the solution was then subjected to hydrothermal treatment in a 100 ml stainless steel autoclave lined with Teflon under autogenous pressure. The autoclave was maintained at 200 °C for 24 h and then cooled to room temperature. The as-fabricated products were collected and washed with ethanol and deionized water, respectively. And then, the products were dried in an oven at 60 °C for 12 h. CdS nanoparticles fabricated by this method was denoted as H-CdS.

Characterization of photocatalysts:

IR spectra were obtained on Bruker TENSOR 27 FT-IR spectrometer with a resolution of 1 cm^{-1}. XRD patterns were collected on X'pert MPD Pro (PANalytical Co.) diffractmeter using Cu-Kα radiation (40 KV, 40 mA). All the scanning electron microscopy (SEM) images were taken on a JEOL JSM 6700F field emission instrument operated at 5.0 kV. Transmission electron microscopy (TEM) images were recorded using a Tecnai T20 microscope, with Energy dispersive spectra (EDS) analysis collected in the meantime. UV-Visible (UV-vis) absorption spectra of the samples were recorded using a HITACHI UV4100 operating between 800 nm to 240 nm.

Photocatalytic H$_2$ production experiment:

The water splitting experiments were carried out under visible light illumination(λ>420 nm). 0.05 g photocatalysts were dispersed in a 50 mL aqueous solution containing 0.1M Na_2S and 0.04 M Na_2SO_3 which acted as sacrificial agents to protect photocatalysts from photo-degradation. The light source was a 500 W Xe lamp(CHF XM500W, Beijing Trusttech Co. Ltd., China), and the light intensity reaching the surface of the reaction solution was tested to be 181 mW·cm^{-2}. The amount of H_2 evolution was determined on a gas chromatography (Agilent Technologies: 6890N) with TCD equipped with a 5A molecular sieve using N_2 as the carrier gas.

RESULTS AND DISCUSSION

FT-IR spectra

Figure 2 shows the FT-IR spectra of the precursor fibers as well as the other calcinated samples under differing temperatures. Curve a represents the PVP/ zinc acetate/ cadmium acetate composite fibers (precursor fibers), and the peaks at about 3200, 2958, 1647, 1568, 1288, 1026, and 671cm^{-1} corresponds to v_{N-H}, v_{C-H}, $v_{C=O}$, v_{C-C}, v_{C-O}, v_{O-H}, and $v_{C=S}$, respectively[25]. The peaks around 399 and 620 cm^{-1} assigned to v_{Zn-O} of ZnO appeared after calcined at 480 °C. When the annealing temperature was up to 600 °C (Fig. 1c), all the peaks that belongs to PVP and CH_3COO^{2-} groups disappeared, whereas the peaks of v_{Zn-O} and SO_4^{2-} (1112cm^{-1}) were greatly enhanced. When the annealing temperature reached to 700 °C (Fig. 1d), only the bands at 399, 620 and 1132 cm^{-1} assigned to ZnO and CdSO$_4$ were left. All the peaks belonging to organic groups disappear. Notably, there always existed three peaks at 1647, 3450, and 2346cm^{-1} in Fig.1 b-d, which can be assigned to hydroxyl water and carbon dioxide absorbed by the fiber samples. These results illustrated that the organic molecules could be removed completely from PVP/zinc acetate/cadmium acetate composite fibers when the calcining temperature was above 480 °C.

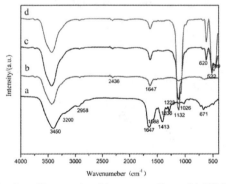

Figure 2. FT-IR spectra of various fiber samples: (a) precursor fibers; (b) 480 °C; (c) 600 °C; (d) 700 °C

XRD Phase Analysis

Figure 3 shows the XRD patterns of various fiber samples annealed under different temperatures. From the patterns, it can be presumed that, after the composite fibers were calcined at 480 °C, they exhibited a mixture of crystalline phase corresponding to hexagonal CdS, hexagonal ZnO and a few Cd_3OSO_4 phases (JCPDS 02-3140). The five reflection peaks (△)appearing at 2θ = 24.83°, 26.48°,

28.16°, 43.78° and 52.13° correspond to the hexagonal CdS crystalline phase (JCPDS 41-1049). The other obvious peaks(□) at 2θ =31.76°, 34.39°, 36.2°, 47.51°, 56.59°, 62.85°, 67.95 °, 69.06°, corresponds to the hexagonal ZnO crystalline phase (JCPDS 36-1451), which can be assigned specifically to the (100), (002), (101), (102), (110), (103), (112) and (202) planes. Notably, when the calcining temperature increased to 600 °C and 700°C, instead of the appearance of hexagonal CdS phase peaks, CdSO$_4$ crystalline phase appeared, which suggested that the crystals transform from CdS to CdSO$_4$ and Cd$_3$O$_2$SO$_4$ phase with the increasion of calcining temperature. This phenomenon is similar to the results published in the ref. 26.

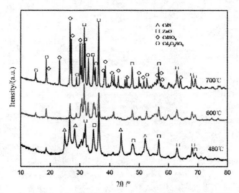

Figure 3.   XRD patterns for various fiber samples treated under different temperatures

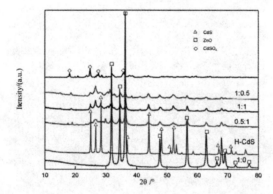

Figure 4. XRD patterns of H-Cd and (ZnO)x/(CdS)y composite fibers calcinated at 480°C (where x and y denote the molar ratio of Zn and Cd precursors in the preparation process)

Figure 4 shows the XRD patterns of H-CdS and the composite fibers with different Zn/Cd ratio which were calcined at 480°C. The patterns demonstrate that ZnO/CdS composite crystalline phase could be obtained at different ratio. What needs to point out is that no CdS characteristic peaks appear in the sample without Zn. The H-CdS sample is hexagonal CdS phase, and the 1:0 sample has extreme high intensity. The possible reason for this is that the bare ZnO has low crystallization temperature. The average grain sizes of the products were calculated by applying the Debye-Scherrer formula, $D=K\lambda/(\beta\cos\theta)$, where λ is the wavelength of the X-ray radiation ($CuK\alpha=0.15406$ nm), K is a constant taken as 0.89, β is the line width at half maximum height of the peak, and θ is diffraction angle. The results are listed in Table 1.

Table1. Grain sizes, diameters and bandgaps of the as-electrospun nanofibers and the H-CdS

Sample(x:y)	Grain size (nm) ZnO/CdS	diameter (nm)	Bandgap(eV)
H-CdS	/ >100	/	2.38
0.5:1	18/12.1	200	2.48
1:1	19.5/17.8	300	2.53
1: 0.5	16.5/10.7	100	2.53
1:0	27.2/	70-120	3.33

SEM Morphology Analysis

SEM photographs of the precursor fibers and the fibers obtained at different temperatures are shown in Figure 5. The collected precursor fibers have a smooth surface and an average diameter of 400 nm as seen in Figure 4a and b. After thermal treatment at 480 °C, the diameter of fibers shrinks to 300 nm and the surface becomes rough because of the pyrolysis, reduction and crystallization. It can be estimated that the length of the nanofibers could reach to several decimeters from Figure 4c. As one can see the red cycle area of the Figure 4d, the products that we got at 480 °C were core/shell-type nanofibers. After the calcining temperature increased to 600 °C (Figure 4e and f), the diameter of the fibers continuously decreases and core/shell nanofibers were replaced by porous hollow fibers because the core material CdS transform to $CdSO_4$ and $Cd_3O_2SO_4$ in the high temperature treatment, which is derived from the XRD results. When the calcining temperature further reached to 700 °C (Figure 4g and h), the continuous fibers exhibit much defections, because the grain size generally increased.

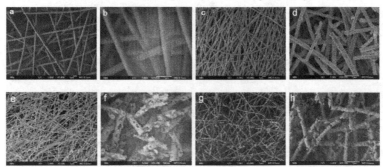

Figure 5. SEM images for various fibers calcined at different temperatures: (a, b) precursor fibers; (c, d) 480 °C; (e,f) 600 °C; (g, h) 700 °C

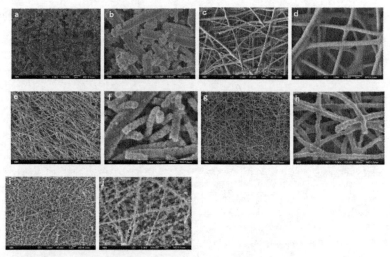

Figure 6. SEM images of H-Cd (a, b); (ZnO)x/(CdS)y composite fibers calcined at 480 ºC (where x and y denote the molar ratio of Zn and Cd precurors in preparation), 0:1(c, d); 0.5:1(e, f); 1:0.5(g, h); 1:0(i, j)

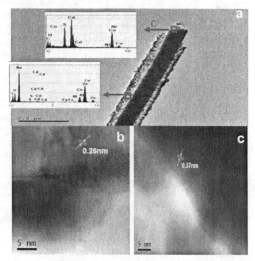

Figure 7. TEM and EDS results for core/shell fiber sample

In order to describe the effect of different ratio of Zn/Cd on the morphology of the fibers, we synthesized various composite fibers with different Zn/Cd ratio (see in fig.5), including 0:1, 0.5:1, 1:1,

1:0.5 and 1:0. It can be found that the core/shell-type nano-fibers only emerge in the 1:1 sample, and the 1:0 sample is nanoparticles-nanofibers mixture structure, which coincided with the explanation we mention above. The H-CdS (fig. 5 a, b) showed that CdS nanorods appear, which are also 1D nanomaterial as same as the nanofibers.

TEM and EDS Analysis

The structure and morphology of the core/shell fibers were further characterized by the transmission electron microscopy. Figure 7 a and b clearly show that the diameter of core fibers was mainly 200 nm and the shell layer was 50 nm, which was in accordance with the SEM observation. The surface of the calcination fibers was rough due to the crystallization of ZnO. A high-resolution image of the CdS/ZnO core/shell nanofibers reveals that the simultaneous presence of crystalline CdS and ZnO crystal lattices in the region of the heterojunction. The fringes observed correspond to the interplanar distances of 0.37 and 0.26 nm, which agree well with the lattice spacing of the (100) and (002) of the hexagonal CdS and ZnO crystal, respectively. The analysis of energy-dispersive spectroscope (EDS) shows the elements of core fibers and shell layer. The results of EDS analysis are presented in Figure 6c, d. It shows that the core fiber is CdS and the shell layer is ZnO. The C, Cu, Fe, Ni and Na elements in the core fiber or shell layer were from the substrate of the equipment or impurities of the fibers.

UV-vis absorption spectra

Figure 8 shows the UV-vis absorption spectra of the H-CdS, $(ZnO)x/(CdS)y$ nanofibers and pure ZnO. In Figure 7, the absorption edges of ZnO and H-CdS nanofibers start at 372 and 520 nm, corresponding to the band gap energies of 3.33 and 2.38 eV[27], respectively. These values are in consistent with the reported bandgaps of ZnO (3.37 eV) and CdS (2.4 eV). As for these samples, an enhanced absorption in the visible light region (at wavelength over 400 nm) was clearly observed with the increase of CdS content. This can be ascribed to the small band gap (2.4 eV) of CdS (seen in Table 1).

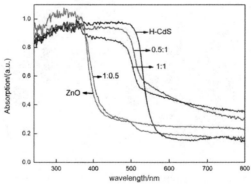

Figure 8. UV-visble absorption spectra of ZnO, $(ZnO)x/(CdS)y$ composite fibers and H-CdS

Combined with the results of XRD, SEM, TEM, and UV-vis spectra, it can be concluded that the formation of the ZnO/CdS heterojunction in ZnO/CdS nanofibers. It is assumed that the ZnO/CdS nanofibers might have promising photocatalytic efficiency.

Photocatalytic activity test

Photocatalytic H_2 evolution of $(ZnO)x/(CdS)y$ varies with the ratio of ZnO to CdS. The composite fibers (seen in Figure 9) were dispersed in an aqueous solution containing SO_3^{2-} and S^{2-} ions as sacrificial reagents under simulated visible light ($\lambda > 420$ nm) irradiation. The photocatalytic H_2 evolution of $(ZnO)x/(CdS)y$ has no regular increase or decrease trend with the change of ZnO to CdS molar ratio, although their optical absorption spectra edges are gradually decreases with the increase of ZnO molar ratio. Among all of the $(ZnO)x/(CdS)y$ photocatalysts, the $(CdS)1/(ZnO)1$ core/shell nanofibers exhibit the highest H_2 evolution of 2100 mmol·g^{-1} for 6 h(354mmolh$^{-1}g^{-1}$), which is twice and 21 times higher than that of ZnO nanoparticles-nanofibers prepared by the electrospinning and CdS prepared by the hydrothermal route, respectively. This result is different from the trend investigated in the reported ZnO/CdS core/shell-type nanorods prepared by the two step route[14]. We believe that the core/shell-type structure plays a key role in improving the photocatalytic activity. This conclusion can be understood as follows: firstly, in this band-gap configuration, when electron-hole pairs are generated by visible light excitation in CdS core, the photoelectrons can swiftly transferred to the conduction band （CB） of the ZnO shell[28, 29], which facilitates the charge separation process of electron-hole pairs. On the other hand, thin ZnO shell is transparent under visible light, the visible light can be harvested by the CdS core at the same time [30]. Furthermore, once electrons diffuse from CB of CdS into the CB of ZnO, the probability of their decay is small because there can be no free holes in ZnO under visible excitation. Secondly, ZnO as the shell layer can protect the CdS core from photocorrosion, and prolong the lifetime of the photocatalysts. In addition, this structure increases the exposed ZnO surface as reduction sites for H_2 evolution. Thirdly, the core/shell-type structure possess large interfacial region, which can improve the charge carrier separation to a large extent. Furthermore, the less lattice mismatch between the ZnO shell and CdS core would result in a more defect-free interfacial region[31], which decreases the carrier trap chance in the interface region, and increases the charge separation efficiency.

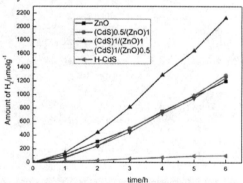

Figure 9. The performance of ZnO, $(ZnO)x/(CdS)y$ composite fibers and H-CdS hydrogen production under visible light

CONCLUSION

For the first time, CdS/ZnO core/shell nanofibers with core fibers mainly 200 nm and the shell layer 50 nm were prepared via the typical (single-nozzle) electrospinning and calcination treatment by using the electrospun thin fibers of PVP/zinc acetate/cadmium acetate as precursor. This route might open a new door to make core/shell nanofibers of inorganic materials. By modifying the parameters of sol-gel or electrospinning process, one could also expect to be able to prepare core/shell nanofibers of inorganic materials with different diameters. The prepared CdS/ZnO core/shell-type nanofibers showed comparatively high visible light photocatalytic H_2-production activity even without any cocatalyst. The unique core/shell-type structure exhibits a significant influence on the visible light photocatalytic and the corresponding highest H_2-production rate is 354 μmol h^{-1} g^{-1}. It is believed that the core/shell-type structure facilitates the charge separation process of electron-hole pairs. Moreover, the ZnO shell layer is transparent under visible light and they can protect the CdS core from photocorrosion, and prolongs the photocatalyst lifetime.

ACKNOWLEDGMENTS

This work was supported by the Fundamental Research Funds for the Central Universities of China (2011JDGZ15), the Specialized Research Fund for the Doctoral Program of Higher Education of China (20090201110005) and Natural science fund of Jiangsu Province, China (SBK201022919). The authors wish to thank Feng Jiangtao for his technical assistance.

REFERENCES

[1] Akihiko Kudo, Yugo Miseki, Heterogeneous photocatalyst materials for water splitting, *Chem. Soc. Rev.*, **38**, 253-78 (2009).

[2] A. Fujishima, K. Honda, Electrochemical Photolysis of Water at a Semiconductor Electrode, *Nature*, **238**, 37-38 (1972).

[3] S. K. Panda, S. Chakrabarti, B. Satpati, et al., Optical and microstructural characterization of CdS-ZnO nanocomposite thin films prepared by sol-gel technique, *J. Phys. D: Appl. Phys.*, **37(4)**, 628-33(2004).

[4] Vasa Parinda, B. P. Singh, Ayyub Pushan, Coherence properties of the photoluminescence from CdS-ZnO nanocomposite thin films, *J. Phys. Condens. Matter*, **17(1)**, 189-97(2005).

[5] Ayyub Pushan, Vasa Parinda, Taneja Praveen, et al., Photoluminescence enhancement in nanocomposite thin films of CdS-ZnO, *J. Appl. Phys.*, **97(10, Pt. 1)**, 104310/1-104310/4(2005).

[6] Du Ning, Zhang Hui, Chen Bingdi, et al., Low-temperature chemical solution route for ZnO based sulfide coaxial nanocables: general synthesis and gas sensor application, *Nanotechnology*, **18(11)**, 115619/1-115619/6 (2007).

[7] X. Q. Meng, D. X. Zhao, J. Y. Zhang, et al., Photoluminescence properties of single crystalline ZnO/CdS core/shell one-dimensional nanostructures, *Mater. Lett.*, **61(16)**, 3535-38(2007).

[8] Irimpan Litty, V. P. N.Nampoori, P. Radhakrishnan, Spectral and nonlinear optical characteristics of nanocomposites of ZnO-CdS, *J. Appl. Phys.*, **103(9)**, 094914/1-094914/8(2008).

[9] V. V.Shvalagin, A. L.Stroyuk, I. E. Kotenko, et al., Photocatalytic formation of porous CdS /ZnO nanospheres and CdS nanotubes, *Theor. Exp. Chem.*, **43(4)**, 229-34(2007).

[10]Tak Youngjo, Kim Hyeyoung, Lee Dongwook, et al., Type-II CdS nanoparticle-ZnO nanowire heterostructure arrays fabricated by a solution process: enhanced photocatalytic activity, *Chem. Commun.*, **38**, 4585-87(2008).

[11]Joo Jinmyoung, Kim Darae, Yun Dong Jin, et al., The fabrication of highly uniform ZnO/CdS core/shell structures using a spin-coating-based successive ion layer adsorption and reaction method, *Ĵanotechnology*, **21(32)**, 325604/1-325604/6(2010).

[12]Lee Wonjoo, Min Sun Ki, Dhas Vivek. Chemical bath deposition of CdS quantum dots on vertically aligned ZnO nanorods for quantum dots-sensitized solar cells, *Electrochem. Commun.*, **11(1)**, 103-6(2009).

[13]Youngjo Tak, Suk Joon Hong, Jae Sung Lee, et al., Fabrication of ZnO/CdS core/shell nanowire arrays for efficient solar energy conversion, *J. Mater. Chem.*, **19**, 5945–51(2009).

[14]Xuewen Wang, Gang Liu, ZhiGang Chen, et al., Enhanced photocatalytic hydrogen evolution by prolonging the lifetimeof carriers in ZnO/CdS heterostructures, *Chem. Commun.*, **23**, 3452–54(2009).

[15]Paromita Kundu, Parag A. Deshpande, Giridhar Madrasb, et al., Nanoscale ZnO/CdS heterostructures with engineered interfaces for high photocatalytic activity under solar radiation, *J. Mater. Chem.*, **21**, 4209-16(2011).

[16]Zhou Xinhong, Shang Chaoqun, Gu Lin,et al., Mesoporous Coaxial Titanium Nitride - Vanadium Nitride Fibers of Core-shell Structures for High-Performance Supercapacitors, *ACS Appl. Mat. Interfaces,* **3(8)**, 3058-63(2011).

[17]Park Jae Young, Choi Sun Woo, Lee Jun Won,et al., Synthesis and gas sensing properties of TiO$_2$-ZnO core-shell nanofibers, *J. Am. Ceram. Soc.*, **92(11)**, 2551-54(2009).

[18]Fragala M. E., Cacciotti I., Aleeva Y., et al., Core-shell Zn-doped TiO$_2$-ZnO nanofibers fabricated via a combination of electrospinning and metal-organic chemical vapour deposition, *CrystEngComm*, **12(11)**, 3858-65(2010).

[19]Luo Wei, Hu Xian Luo, Sun Yong Ming, et al., Electrospinning of carbon-coated MoO$_2$ nanofibers with enhanced lithium-storage properties, *PCCP*, **13(37)**, 16735-40(2011).

[20]Cao Tieping, Li Yuejun, Wang Changhua, et al., Three-dimensional hierarchical CeO$_2$ nanowalls/TiO$_2$ nanofibers heterostructure and its high photocatalytic performance, *J. Sol-Gel Sci. Technol.*, **55(1)**, 105-110(2010).

[21]Nagamine Shinsuke, Ishimaru Shingo, Taki Kentaro,et al. Fabrication of carbon-core/TiO$_2$-sheath nanofibers by carbonization of poly(vinyl alcohol)/ TiO$_2$ composite nanofibers prepared via electrospinning and an interfacial sol-gel reaction, *Mater. Lett.*, **65(19-20)**, 3027-29(2011).

[22]Zhang Peng, Shao Changlu, Zhang Zhenyi, et al., TiO$_2$@carbon core/shell nanofibers: controllable preparation and enhanced visible photocatalytic properties, *Ĵanoscale*, **3(7)**, 2943-2949(2011).

[23]Lin Dandan, Wu Hui, Zhang Rui, et al., Facile synthesis of heterostructured ZnO-ZnS nanocables and enhanced photocatalytic activity, *J. Am. Ceram. Soc.*, 2010, **93(10)**, 3384-89.

[24]Choi Sun Woo, Park Jae Young, Kim Sang Sub, Synthesis of SnO$_2$-ZnO core-shell nanofibers via a novel two-step process and their gas sensing properties, *Ĵanotechnology*, **20(46)**, 465603/1-465603/6(2009).

[25]Y Zhang, J Li, Q Li, et al., Preparation of CeO$_2$-ZrO$_2$ ceramic fibers by electrospinning, *J. Colloid Interface Sci.*, **307 (2)**, 567-71 (2007).

[26]Éder T.G. Cavalheiroa, Massao Ionashirob, Glimaldo Marinoc. et al., The Effect of the Aminic

Substituent on the Thermal Decomposition of Cyclic Dithiocarbamates, *J. Braz. Chem. Soc.*, **10(1)**, 65-75(1999).

[27]Zhenyi Zhang, Changlu Shao, Xinghua Li, et al. Electrospun Nanofibers of ZnO-SnO$_2$ Heterojunction with High Photocatalytic Activity, *J. Phys. Chem. C*, **114 (17)**, 7920-7925(2010).

[28]Spanhel Lubomir, Weller Horst, Henglein Arnim, et al. Photochemistry of semiconductor colloids. 22. Electron ejection from illuminated cadmium sulfide into attached titanium and zinc oxide particles, J. Am. Chem. Soc., **109(22)**, 6632-35(1987).

[29]Hotchandani Surat, V. Kamat Prashant, Charge-transfer processes in coupled semiconductor systems. photochemistry and photoelectrochemistry of the colloidal cadmium sulfide-zinc oxide system, *J. Phys. Chem.*, **96(16)**, 6834-39(1992).

[30]Minmin Shi, Xiaowei Pan, Weiming Qiua. Si/ZnO core-shell nanowire arrays for photoelectrochemical water splitting, *Int. J. Hydrogen Energy*, **36(23)**, 15153-59(2011).

[31]Yoon Myung, Dong Myung Jang, Tae Kwang Sung et al. Composition-Tuned ZnO-CdSSe Core-Shell Nanowire Arrays, *Jano Lett.* , **4 (7)**, 3789-3800(2010).

Author Index